MW00444993

Modern Accident Investigation and Analysis

SECOND EDITION

Modern Accident Investigation and Analysis

SECOND EDITION

Ted S. Ferry
University of Southern California

A WILEY-INTERSCIENCE PUBLICATION

JOHN WILEY & SONS

New York • Chichester • Brisbane • Toronto • Singapore

Library of Congress Cataloging in Publication Data:

Ferry, Ted S.
Modern accident investigation and analysis/Ted S. Ferry—2nd ed.
p. cm.
"A Wiley–Interscience publication."
Bibliography: p.
Includes index.
ISBN 0-471-62481-0
1. Industrial accidents—Investigation. I. Title.
HD7262.25.F48 1988 87-34027
658.3'82—dc19 CIP

Printed in the United States of America

20 19 18 17 16

Foreword

"Every accident, no matter how minor, is a failure of organization." This quotation delivered in 1953 by Professor Kenneth R. Andrews of the Harvard Graduate School of Business Administration, carries the implication that an accident is a reflection on management's ability to manage, except for acts of God. Accident prevention is a management function. Even minor incidents are symptoms of management incompetence that may result in a major loss.

Accident prevention depends to a large degree on lessons learned from accident investigation. However, it is the rare manager who has acquired the intelligence necessary to evaluate skill and objectivity in conducting the in-depth search for the cause of an injury or damage to equipment. This book should endow the manager with the facility to review with considerable confidence the accident investigation procedures and reports submitted by his staff. He does this with balance sheets, sales charts, and production plans; why not with profits and losses that result from his safety programs or accidental losses? It is all part of organizational performance.

Accidents can be extremely costly, as well as having undesirable impacts on organizational morale, on organizational image, on productivity, and on fulfilling commitments. They have tangible and intangible effects on "the bottom line."

The manager, or chief executive officer if you will, should understand that an objective investigation to determine the root causes of an accident is both a science and an art. As a science it demands special techniques and disciplines that are unfolded in this book. It is an art because of the background of experience, even an acquired intuition, needed to bring together the information and analyze it properly for a productive conclusion, letting the chips fall where they may. The root cause may well be traced to a management decision in the distant past.

While the manager depends on specialists to conduct the investigation, he should be able to intelligently judge their performance. This book will provide him with an excellent start in that direction. It should also serve as a primer for those interested in making a career of accident investigation.

JEROME LEDERER
President Emeritus
Flight Safety Foundation
and Retired Director of Safety
NASA

Preface

This book covers the techniques needed to investigate accidents or review the work of those who do. There are relatively few publications on the subject, and most deal with highly specialized areas, such as traffic or aviation. Until the appearance of the first edition of *Modern Accident Investigation and Analysis,* it was difficult to recommend an appropriate general text for the hundreds of lay persons and students who needed one. Like the earlier edition, this volume is a guide for those who do not have an experienced investigator at hand but seek their own counsel and solutions. It will also provide the seasoned investigator with a resource that will both furnish refreshing new ideas and serve as a reference.

The main purposes of the book are to:

1. Provide enough information to investigate those events called accidents to an extent that will ensure identifying causal factors so they can be corrected to prevent more accidents, and
2. Ensure, through good investigation, corrective action that will improve operation of the organization.

This secondary purpose is not covered in depth but the techniques presented within will make it practical.

Two words should be noted: manager and mishap. "Manager" in this book means someone in the organizational structure, in either a line or staff position, who is involved in operations and has a stake in efficient management. "Mishap" is used to describe any of those unexpected, undesired events that cause a loss of resources: time, lives, money, hardware, or whatever. I prefer the term "mishap" over "accident," which defies standardization and common acceptance. "Mishap" serves better as an overall descriptive term that is not so apt to be at odds with the reader's particular definition of accident, incident, and similar words.

This book is for those who need a knowledge of mishap investigation for occasional investigative duties and who must review mishap reports to ensure proper corrective actions and further loss of resources. A wide variety of techniques are covered in suitable depth for competent, if not expert, use. Expertise comes only from extensive hands-on experience over a period of time. The investigator's knowledge of the information contained here will ensure a competent investigation that can serve management in many ways.

Most mishap investigations do not call for in-depth skills. In practice, regulatory requirements often mandate only a cursory study of the facts. A modest level of training is sufficient. Unfortunately, mishap investigation merely to satisfy the regulations usually does little to prevent future mishaps. Causal factors that reveal inefficient operations and unacceptable/ ways of operating are seldom isolated. This book provides the background and knowledge to make mishap investigation a profitable and practical management tool.

Management is emphasized over first line supervision because investigative skills have a low priority for most supervisors. The supervisor often does not have the time, knowledge, and skill to do a thorough investigative job. These limitations are acknowledged here in order to guide and assist him. When an investigation is complete, management must be able to recognize the uncovered causes, weight them, and then take the appropriate corrective action.

This book covers many investigative techniques and tools. No one technique is a magic answer to mishap investigation. Critical readers will instinctively seek out the path and techniques that best serve their purposes.

The arrangement of the contents is straightforward. Background material and a discussion of the need for investigation are presented in Chapter 1. Chapters 2 and 3 deal with preparing for, organizing, and managing the investigation. Chapters 4, 5, and 6 cover the interacting roles of human beings, the complete environment, and materials. Chapter 7 further develops the interaction process with a systems approach to investigation. In Part 3, Chapter 8–15, various analytical techniques are separately reviewed in enough detail to be applied in practice. The proper reporting of mishaps, from the minimum required by law to that needed to document for liability, compensation, and detailed corrective action, is thoroughly treated in Chapters 16 and 18. Chapter 17 is a new chapter and covers the 12 steps of mishap investigation. Newer approaches and future techniques of investigation are overviewed in Chapter 19, the last chapter in the book.

If the aim of the book were to be summarized in one sentence, it would be: To allow us to investigate mishaps with confidence, capability, and credibility; make recommendations to prevent similar mishaps from occuring in the future; and contribute to the overall effectiveness of the organization.

This material will take much of the mystery out of investigation and encourage a better understanding of mishap causes.

Readers will find this revision has changed considerably from the original book. Each sentence has been reviewed and often revised, new material has been added, and new and more current references have been added to reflect many advances in the few years since the first edition.

TED FERRY

Newbury Park, California
December, 1987

Acknowledgments

The people who assisted in making the first edition of this book possible are many. Appreciation is due to the those early contributors to the material included here: Lewis Bass, Ludwig Benner, Jr., John Dreher (deceased), Jerome Lederer, Harold Roland, D. A. Weaver, Richard Wood, and countless graduate students at the University of Southern California. D. A. Weaver and Richard Wood have reviewed their earlier contributions and suggested significant changes to this revision.

Since the appearance of the first edition, students and practitioners alike have contributed ideas and corrections to this revision. Often as not, their contribution was made unknowingly, during a conversation or a speech or through a letter. Frequently they made reference to an information source that turned out to be a small gold mine. The revision was blessed with the support of the Safety Sciences Department at the University of Southern California. Faculty contributors to this edition were Professors Charles Dole, P. A. Hancock, Monsour Rahimi, and Richard Wood, while administrative support came from Natalie Constanza, Jill Floyd, Connie Henighan, and Karen Wong. Other valuable support and suggestions came from Jerome Lederer, formerly of NASA and the Flight Safety Foundation, John O'Toole of Cal OSHA, and Harry Borowka of the Merritt Company.

Also noteworthy is the continuous cooperation of professional societies and organizations who freely made their material available for both editions of *Modern Accident Investigation*.

T.S.F.

Contents

Modern Accident Investigation and Analysis

SECOND EDITION

PART ONE

The Who, What, Why, and When of Investigation

The first part of the book primes the reader for mishap investigation. Three chapters lead into the serious business of investigation through a discussion of the need for investigation and an examination of who has a stake in it. This is followed by coverage of the preparation that makes an efficient investigation possible. In Chapter 3 treatment of the investigation itself gets under way as we learn the first important steps to take at the scene of a mishap.

CHAPTER ONE

The Need for Investigation

INTRODUCTION

Investigation is a small part of safety and health program administration, yet it is essential if management is to profit from the errors and omissions that have resulted in any unintended loss of resources. The magnitude of many problems facing effective and efficient investigation often makes them appear to the beyond practical solution, but there has been progress in recent years.

Mishaps may be investigated by parties ranging from untrained persons with limited resources working alone to large investigative teams of experts with nearly unlimited resources. Most mishaps are investigated by persons without any investigative background who have no particular approach to the task. They usually have minimum resources to meet minimum company or government regulations. This has placed us in the situation of getting little benefit from most investigations. The remarks of Edwards size up the situation[1]:

> The haphazard nature of accident investigation and analysis provides none of the factors for a base for constructive and positive accident prevention policy, although much time and effort is given to collecting information which is not put to constructive use.
>
> Accident reporting systems have not been designed as information systems, but have been grown in a relatively unplanned way. We need a method of reporting accidents capable of providing an accurate and effective basis for line management decision making.

And a comment from another source[2] partly covers the situation:

> The poor quality of data has influenced our analysis (of mishap investigations) from the point of view that some analysis that could have been taken were not and some which were conducted produced conclusions which were suspicious.

The scope of an investigative report should take in three distinct areas: (1) what takes place before a mishap occurs, seen as a preparation phase; (2) the gathering of factual information surrounding the mishap, seen as information development; and (3) the often neglected activity after all causal information is developed, viewed here as the action phase.

OBJECTIVES OF INVESTIGATION

Several valid investigation objectives are listed below. The list could be greatly expanded. The rationale for each objective is usually self-evident.

1. Reduce danger to employees and susceptible public.
2. Prevent company resource losses.
3. Prevent further mishaps.
4. Respond to management needs.
5. Prevent loss of trained personnel.
6. Develop costing information.
7. Improve operating efficiency.
8. Provide answers to address anticipated public concern.
9. Define operating errors.
10. Define management errors.
11. Satisfy company rules.
12. Reduce work process disruption.
13. Provide protection against litigation.
14. Satisfy insurance requirements.
15. Improve company product.
16. Educate supervisors.
17. Develop cost information.
18. Anticipate government interest.
19. Identify violations of company procedures.
20. Comply with workers' compensation rules.
21. Protect against litigation.
22. Satisfy regulatory requirements.
23. Educate management and staff.
24. Research purposes.
25. Improve quality control and reliability.
26. Isolate design deficiencies.
27. Satisfy news media.

The order of listing is not important. Commonly voiced objectives that lack sound reasoning are not included. Samples of those are (1) do it for the sake of appearances, (2) justify safety manager's job, and (3) "hang someone to set an example."

Investigating to determine liability is quite different from the objective of defining management errors. While it may seem that one good investigation would serve all purposes, for legal reasons and practical expenditure of resources this is not feasible. It is true that the more complete and in-depth an investigation is, the more likely it is to serve more objectives. However, it are still present; the mistakes of management and supervision are there; the operator and design deficiencies still exist. The ideal approach is to investigate all undesirable and out-of-the-ordinary events that keep us from doing our job efficiently. They deserve the same careful scrutiny that a loss of life or serious property damage would bring. I have therefore settled on the term "mishaps" for these events, since often they cannot technically be called accidents. Others might call them near misses, close shaves, or minor incidents.

In some establishments these close shaves or near misses are looked into with the same spirit of inquiry as a major event. Unfortunately, few of us have the time to look in detail at every event that could have been an accident except by definition. For one thing, we may not have the personnel or other resources to justify such an effort, but experience can guide us. An alert manager knows the mishap potential when a goof-up has happened too many times or when there could have been serious injury or substantial property loss. If the potential for significant loss is present, then there should be an investigation. The minor mishap, then, often deserves to be investigated with the same spirit as if a major mishap had occurred. The benefits from finding causal factors before they cause a recordable mishap offset the resource investment. It is quicker and easier to investigate an event not complicated by injury, damage, potential lawsuit, and the pressures of time than to wait until a recordable event has happened and then investigate.

One expects a serious mishap to be investigated in a timely fashion, while a minor event may not even require a formal report. Realize, though, that the minor event can be a sign of serious trouble. It is a clue that a "real" accident with damage and injury is on the way. The minor mishap, properly investigated, can furnish information that can be used to prevent the serious mishap.

We should investigate any mishap that has the potential for giving us causal factors that also precede major reportable events. When resources do not permit a thorough investigation of minor mishaps, selected sampling and investigation of incidents with great potential for harm can produce good results. We need not wait for a reportable accident to find out what is wrong with the organization.

To see what should be investigated, we must also know how much to investigate. That is, when, for investigation purposes, did the mishap begin and end? As an example, suppose that a forklift driver comes to work in an alcoholic state and, in the process of driving his defective forklift, has a mishap. He is then thrown from the forklift into moving machinery and is badly hurt.

For the purposes of many investigations we might only be required to determine his condition when driving the forklift and not be concerned with what happened after he was thrown from the machine. We might say that the investigation should consider only his operation of the forklift that resulted in the mishap. Relevant factors might include his sobriety, his operation of the forklift, the faulty condition of the forklift, or his operating error that led to the mishap. We might expand a bit to include the movement of the forklift and where it came to rest. A thorough investigation would go farther to determine if, why, and how he came to work in an alcoholic state. Had he done so before? If so, why was it permitted? We would look at the reasons for this forklift being defective and its involvement in this particular mishap. We would be concerned about the cause of the operator's injuries and what made that factor possible. We cannot afford to investigate until every possible causal factor is tracked down and the correction of every possible cause of the event is carried out. There are few provisions for such detail in the average business or industry, yet preventing similar events calls for as detailed an investigation as possible.

Sometimes the investigation of such a mishap starts with the first movement of the machine and ends when it stops moving. At other times a mishap injury is not considered important except as an unfortunate by-product of the mishap. At still other times the traffic mishap investigation is carried back to the driver's alcoholic state, his psychological basis for being drunk, and any other reasons bearing on his problem. This book also recommends the investigation of any management function. If the investigator has the time, resources, and permission to investigate thoroughly, the rule is this: *Investigate causal factors until it is no longer feasible to correct the deficiencies uncovered, and make specific recommendations for corrective action on each deficiency.*

It is important to know just what is to be investigated, exactly when the event started and when it stopped for investigation purposes. You must know how far and how deeply you are allowed to investigate. This can be part of the planning process and should be decided soon after arrival at the investigation scene. Another approach is to investigate all mishaps as required by law and company rules and do every fifth or tenth mishap in detail from start to finish.

WHO ARE THE STAKEHOLDERS?

This book is directed at all those who have a personal stake in investigation, either in conducting investigations or in reviewing them for corrective action. Who are these stakeholders? For industrial mishaps, most investigations are carried out by the supervisor/foreman, line manager, safety professional, staff manager, or special committee. Sometimes a special group or individual, by virtue of the consequences or uniqueness of the event, has been asked to take part. This includes the specialized investigator, who, because of his unique skills and training, spends a large part of his time investigating mishaps. They all have a heavy stake in the investigation.

Supervisor/Foreman

The supervisor/foreman is closest to the action. The mishap takes place in his domain, and he is most likely to be the one who investigates it. When the operation is large and has several mishaps each year, we can expect others to help or have investigative duties. In a large operation, the supervisor may look into only those mishaps that result in minor injuries or a small loss of resources. In any event, as the supervisor of the operation he will furnish much of the required data, and he may prepare a part of the necessary reporting forms.

The supervisor/foreman seldom has specialized investigative training. There may be many excuses for this lack of training, but few if any valid reasons. If it is the supervisor's duty to investigate, he has every right to expect management to prepare him for the task.

It is reasonable to expect the supervisor/foreman to investigate mishaps in his department. His people and machines are involved. His department orders were being carried out when the mishap occurred. He knows more about the operations and the people than anyone else. He will be charged with the mishap in his area and may be demoted or receive a bonus on the basis of the mishap investigation.

Yet the same reasons for having the supervisor/foreman make the investigations are also the reasons he should not do it. His reputation is on the line. There are bound to be causes found that will, in some way, reflect on his operating method. His closeness to the situation may preclude an open and unbiased approach to the supervisor-caused elements. The more thorough the investigation, the more likely the supervisor will be implicated as contributing to the event. It is hard to be objective when you are blowing the whistle on yourself. While it should never be the objective to hang someone, there cannot help but be some stigma when one's own operation is involved. It requires all the fortitude and objectivity a supervisor can muster to conduct a fair investigation of his own department.

Regardless of who investigates, the supervisor/foreman will interact with the findings. The investigator must turn in a viable report with firm findings and recommend specific corrective actions if similar events are to be prevented from happening in the future. The supervisor/foreman will have to carry out the recommendations. He knows the people involved and the peculiarities of the operation. He is able to communicate directly on a regular basis to ensure corrective action. He knows how to get information from his group, speaks their language, and has their trust. He is in a better position than anyone else to prevent a recurrence. The supervisor/foreman, then, is a major stakeholder in the smooth and efficient operation of the department. The correction of hazarous conditions and operating errors found by a thorough investigation is essential to the welfare of the operation. It ensures increased production time, reduced operating costs, and his continued control of the operation.

Line Manager

The line manager is a person in the management chain who has line authority and is at least a level above the first-line supervisor. He deals directly with the supervisor and foreman instead of the workers; he is part of middle management. The line manager's interest stems from the same background as the supervisor's. At a higher level it is his operation and his people who are involved. People and functions below him are involved, and overall he can act to correct deficiencies found in the investigation. It is to his advantage to have mishaps promptly investigated and corrective actions quickly taken. He may review the supervisor's investigation reports and must recognize when a good job has been done, when the findings are credible and form a basis for action. He, as much as the supervisor, has his reputation riding on a good investigation.

Safety Professional

In any company large enough to have a full-time safety professional, this person should be the company expert in mishap investigation. While supervisors may conduct most of the investigations, the safety professional will review the reports for adequacy and thoroughness. Most important, he will see that appropriate corrective action is taken. As Kuhlman[7] says, the safety professional's role in evaluating investigations permits him to carry out his advisory role to top management. While the ideal situation may keep the safety manager in his evaluation role, there are also times when he should conduct the investigation.

The safety professional is more likely than anyone else in the company to have investigative skills, acquired from reviewing in detail the company mishap reports or from special training taken as part of his professional development. If the supervisors receive special investigative training, the safety professional will often be the one who arranges for this training and may be their instructor. This requires a higher degree of investigative skill than that needed by other people in the organization. Certainly, his skill should be a level above that of the average supervisor. This skill should not be wasted. Where the supervisor is not given time for investigative duties, we can most often expect the safety person to be responsible for them.

When it appears that the supervisor may not be equipped to deal with a mishap, the safety person should make the investigation. Reasons for this range from high public interest potential, to a series of similar mishaps, a mishap that requires skills beyond that of the supervisor, or perhaps a mishap with high potential for liability actions.

Another reason for having the safety person do the investigation is that he does not normally have a vested interest in the outcome. He has no special interest in the affected department and can assess facts and information without bias, prejudice, and the pressures of responsibility for the operation. This advantage may be reduced when workers fear he may be too management-oriented or is investigating his own competence.

Staff Managers

Staff managers are those at the middle-management level who are in a staff advisory capacity, for example, the personnel director, medical director, labor relations manager, purchasing agent, and supply-logistics manager. These persons are seldom line managers, but a good investigation often finds that their functions contributed to the mishap as causal factors. When this is the case, they then have a point in common with the investigator and should be familiar with the investigative process. For example, if we find mishaps happen because of a worker's physical inability to handle the job, it may be necessary to work with the personnel and medical directors in reviewing the worker selection process. Should faulty material or equipment be involved, we might work with the purchasing agent.

As they are far removed from the front line of first-level supervision, there may be a tendency to neglect the staff managers. They can, however, be indispensable in resolving mishap causal factors and may be subject matter experts in the investigation.

Committee Investigations

We classify committee actions into two groups: (1) general or standing committees that have investigation of mishaps as one of their functions and (2) special investigative committees. The general safety committee in an organization often has the task of investigating mishaps when no one has been trained to do the job. In a smaller company, mishap investigation falls naturally within the purview of such a committee. As a standing group whose members serve for extended periods, the committee gains experience based on plantwide mishaps and through investigation training. Since they operate throughout the company, they conduct more investigations than if they were confined to one department.

The general safety committee can bring the knowledge of people with diverse backgrounds to bear on a specific problem associated with the mishap. They can make better investigations and better recommendations. Representing not only the involved department but the company as a whole, they bring parties with and without vested interests together to work on the problem. Such a committee, if representing the organization, tends to have its findings well received as opposed to findings from a member of management or staff.

Special investigation committees are often appointed for serious mishaps.

Their combined skills can be brought into the department and provide in-depth expertise for action where public concern, workers' compensation, or liability can be an issue. Safety committees tend to have their findings better received where fellow workers or supervisory aspects are causal factors. Committee findings may also receive better acceptance when the investigation results are made public.

Committee action has the disadvantage of taking several people away from their normal work to do what some consider the supervisor's job. Sometimes committee members will resent the system, feeling that it gives them the unpleasant task of finding fault with their fellow workers. There is also danger of an unwise compromise being made to reach a consensus.

Other Investigators

A company may not need to do its own investigations. Their insurance carrier may send its own specialists to investigate all or selected mishaps. Sometimes government groups will carry out an investigation, and facts needed for other purposes can often be taken from their reports. The manager who is sincerely interested in mishap prevention may find that the insurance carrier or government investigation is not good enough, and may conduct an in-house investigation.

INVESTIGATOR ROLES

Technical and managerial competence of the investigator does not, in itself, ensure integrity. A technically and managerially well-qualified investigator may unwittingly bias an investigation when he fails to recognize attempts to influence the scope of the investigation.

Some investigators come to the investigation wearing three hats: the fact-finding hat, the determining cause hat, and the hat of an advocate. The third hat is unprofessional and uncalled for in an investigation. In any event, that often determines the approach used.

SUMMARY

Mishaps cause such a drain on our national resources that we must continually seek ways to prevent them. The accidental loss of resources can be greatly reduced through careful investigation. Mishaps have far-ranging implications beyond the direct costs and immediate injuries. Several persons beyond the supervisor and operator are stakeholders. These can include all levels of management and perhaps the public or government. A variety of people can be involved in the investigative process from within or outside the company. They range from the supervisor or special committee to the safety professional and outside experts.

QUESTIONS FOR REVIEW AND EVALUATION

1 Determine how many worker-days of work were lost due to accidents in the United States last year.

2 Find the estimated dollar loss of accidents in the United States last year.

3 Do you agree with the author that it is the task of the investigator to uncover management deficiencies leading to a mishap? Discuss.

4 Give three examples of stakeholders not noted in this chapter.

5 Living where you do, would you send a supervisor to receive special training in accident investigation in another state? Name any organization that gives investigation training.

QUESTIONS FOR FURTHER STUDY

1 Refer to the book *Safety Management* by Grimaldi and Simonds. What is their formula for figuring the cost of accidents?

2 Design an experiment or project that would enable you to determine whether differences in morale can be detected after a mishap in an organization.

3 Explain why the loss of two workers in a factory accident is likely to draw less attention than a minor nuclear incident in which no one is hurt and no property is damaged.

4 Explain why your organization could or could not investigate all mishaps including those not reportable as accidents.

5 Explain why you think the supervisor/foreman should or should not investigate mishaps in his department.

REFERENCES

1 Edwards, Mary, "The Design of an Accident Investigation Procedure," *Applied Ergonomics,* June 1981, Vol. 12, No. 2, pp. 111–116.

2 Drury, Collin G., "Human Factors in Consumer Product Investigation," *Human Factors,* 1983, Vol. 25, No. 3, pp. 329–342.

3 *Accident Facts,* 1986 ed., National Safety Council, Chicago.

4 Ferry, Ted S., "A Review of the Literature of Indirect Costs," *Proceedings: 1979*

Professional Development Conference, ASSE, Don Eckenfelder, Ed., American Society of Safety Engineers, Chicago, 1978, pp. 99–112.

5 Ferry, Ted., "The Price of Safety Demands More than System Safety Expertise," *Proceedings of the 1979 Systems Safety Seminar*, Systems Safety Society, San Francisco, 1979.

6 Root, Norman, *What Every Employer Should Know About OSHA Recordkeeping*, Bureau of Labor Statistics, Department of Labor, Washington, DC, November 1978.

7 Kuhlman, Raymond L., *Professional Accident Investigation*, Institute Press, Loganville, GA, 1977.

CHAPTER TWO

Preparing for Mishaps

Some organizations will never have a mishap that warrants a thorough investigation. Others may have several in a month. In either case we must be ready for mishaps and the investigation that follows. Having a mishap plan lends order to a situation that can quickly get out of hand and upset an orderly process. A viable plan not only aids the investigation but also helps to protect lives and property. It is too late to prepare after a mishap occcurs, so a mishap plan should be prepared ahead of time.

PLANNING FOR MISHAPS

"Planning for mishaps" is not a witty phrase telling us to plan on having mishaps. It refers to having a plan of action ready if we do have a mishap. This plan allows us to proceed in an orderly manner when the unexpected and sometimes disastrous event happens.

Our premishap planning document (we call it a mishap plan) is often called a preaccident plan. It can range from a one-page handwritten sheet nailed on a fence post to a detailed document that covers everything that might be needed in a large-scale investigation. The ideal plan lies somewhere between these two extremes. It must provide clear, concise instructions on what to do and when to do it. It should provide general guidelines on how to proceed. At the least the plan should provide for the following:

- Having and maintaining an alarm system.
- Saving lives.
- Protecting lives and property from more loss.
- Assuring a timely investigation.

The introduction to a mishap plan should say a few words about who the plan applies to and when it should be used. It may be clear to you what event is to be investigated, but the average worker or supervisor will be waiting for guidance when he should be taking action at the mishap scene. There is little room for error when immediate action is needed to save lives or prevent more injuries. Everyone must know what to do when a mishap occurs.

The Alarm System

When a mishap occurs, several people need to be notified at once. Exactly who these people are depends on the situation.

Those near a mishap area must be made aware that a mishap has taken place so they can be alert to an unusual and perhaps dangerous situation. Quick notification through a primary alarm system is needed for those who save lives and protect property—the fire, rescue, and medical teams. After the lifesaving effort is under way, others should be informed through a secondary system. These include the supervisor, manager, on-call investigator, public affairs person, and safety manager.

The primary alarm used to indicate the occurrence of a mishap must have a distinctive warning sound that cannot be confused with anything else. It should get immediate attention from fire, rescue, and medical teams. This is critical in high-noise areas such as where the noise of machinery must be overcome. Where ambient noise is muted, as in a hospital, an easily seen warning light may augment the sound system. The alarm can range from a struck section of resonant pipe to a loud claxon. This alerts the immediate area and is followed or reinforced by announcements over an intercom, radio, or special telephone system. A priority wire (special telephone) system is probably the best and most dependable way to pass the word on a mishap. Such a system does not alarm those who do not need to be involved, thus reducing confusion.

Mishap Plan Details

A mishap plan has key information on what to do in case of a mishap. It includes the names and telephone numbers of persons to be called, usually in a distinctive format on the front cover or first page. The plan assigns specific tasks so that people know what to do and how to go about it. It should be coordinated with other plans such as fire, evacuation, or earthquake plans.

The mishap plan may provide concise information on who is to investigate a mishap, his exact obligations, and the duties of those who support him. It should identify the person to whom a report is to be sent and specify when it is to be sent. A special form is often included, with directions for its use and blanks to be filled in with essential information pertaining to the mishap.

The works manager should be informed of all aspects of the mishap and the investigation. This duty properly falls to the safety director and should be covered in the plan. It is the works manager who will be called by the press, television, and radio news departments. Few things upset a manager more than being told by the news media that a mishap has occurred in his bailiwick when he has no idea what has happened. Reporters have a right to know what has happened. In the absence of facts furnished by someone else, they will print their own version of what happened. The ideal situation is to have the public affairs person deal with the media. This allows a specialist to do his job, and the manager can carry on with running the operation. The safety director is best able to keep them both fully informed. Information should be passed on daily until the matter is no longer news and the manager indicates the pressure is off.

With so many details to consider when a mishap occurs, it is easy for the plan to become cumbersome and hard to keep current. Such a plan is harder to work with and harder to keep abreast of for those who have few mishaps but must know what to do if one happens. It should be concise and easy to grasp. Keep it lean and clean. Nothing ruins a mishap plan quicker than trying to provide for everything. When well designed, the plan can be a checklist to ensure that every essential action is taken. Some plans take the forms of checklists with colored, tabbed pages for easy finding of tasks and responsibilities.

Proper mishap plan design helps to ensure that all parties take quick, proper action when a mishap occurs. The timeliness and quality of the investigation directly reflect the degree of advance planning and the currency of the plan. It is impossible to provide for all eventualities. Improvisation will be needed. Flexibility is essential. The main thing is to get essential actions under way in a timely fashion.

INVESTIGATOR TRAINING

Some specialized training can be found for investigators needing a high degree of sophisticated knowledge (such as aviation), but there are few places that give training in general mishap investigation. A few organizations such as the National Safety Council, the International Loss Control Institute, and the International Safety Academy offer some opportunities. However, they do not meet the needs of hundreds of thousands of businesses and industries for in-house or near-house training, a half-day or full day at a time. The result is a great lack of expertise at the operating levels of business and industry. It is too late for training when the mishap occurs.

The few books on specialized investigation areas are mostly manuals from government agencies or their contractors. Some safety books devote a few words to the subject, seldom as much as a chapter. Few college-level courses address the subject for an entire semester. A handful of films, mostly outdated, cover the subject. Some large insurance carriers have in-house training programs for their safety staff, but these are not often open to the public. The military have long had investigation training programs for their personnel. Many experts count this as their starting point in the field. Special training in aircraft and vehicle mishap investigation is given by schools such as the University of Southern California and Northwestern University. While open to the public, these courses are highly specialized.

Most training is done in-house by the safety manager, who has gained his background through trial and error over an extended period. For the most part, industrial mishaps are looked after by those who could not and do not have to perform with expertise. Yet good investigation is a key to preventing more mishaps. We cannot prevent mishaps without knowing all their causes and their interactions any more than we can prevent illnesses without a thorough knowledge of what causes them. Fortunately, much has been written in articles in safety journals and trade publications. The serious investigator can find good guidance by going through the writings of the last several years.

An ideal solution is to provide each person likely to investigate mishaps with enough knowledge to proceed proficiently with simple or minor investigations. Because there are many more of these types of mishaps, it provides ample opportunity learn, build experience, and better the organization's operation.

INVESTIGATION KITS

The individual supervisor or foreman does not investigate enough mishaps to justify being supplied with a professional kit. Even when supervisors do most of the investigations the kit is usually kept in the safety office. This points out one problem of getting good supervisory investigations. The supervisor cannot justify having a kit for himself and yet he must conduct the investigations. In any event, at least one kit should be available for ready use. Experience has shown that an investigation kit should include the following items:

- Camera and film.
- Clipboard, paper, and pencil.
- Copy of regulations or standard operating procedures.
- Magnifying glass (5× or 10×).
- Report forms.
- Sturdy gloves.
- High-visibility tapes.
- First aid kit.
- Cassette recorder and spare cassettes.
- Graph paper.
- Ruler and tape measure.
- Identification tags.
- Scotch tape.
- Specimen containers.
- Compass.

More complete lists can be found in several places.[1] Sometimes it is appropriate for the kit to contain a few tools. In some places devices to measure noise, gases, and light may be needed. The kit components can be carried in a specially fitted briefcase or camera case and should be clearly marked for easy identification.

Where one person has the principal investigation duty, other arrangements may be appropriate. If sophisticated investigations are to be performed, sometimes quite a distance from home base and perhaps in remote areas, the kit will be more complex. Since the investigator will be required to operate without company support, he will in effect have to live out of his briefcase. Everything that can possibly be needed to investigate should be in it. Travel to distant sites requires provisions for obtaining funds, passports, and transportation and for securing on-site assistance. This must be provided for in advance, so that the investigation is not bogged down by a minor oversight.

Kits should be completely checked periodically and refurbished as needed. Follow a schedule and use a checklist.

PRIORITIES

There is no question about the first priority at a mishap site: Save lives and prevent more injury and property loss. There are procedures for this that apply in most mishaps and lend some order to the situation. The investigator will seldom be the first at a scene unless he is a member of a uniformed state or municipal force. However, since he is an "expert," he can be expected to advise on how these matters should be handled. The California Highway Patrol[2] has two recommendations that should always be kept in mind: First, arrive safely at the scene. Nothing can justify being in such a hurry that another mishap occurs. Second, observe the overall scene on arrival, and begin the planning process.

Anyone responding to a mishap tends to be in a hurry, often with good reason, but your desire to respond quickly should not exclude concern for your own or others' safety. Even the plant fire brigade, which must act immediately, cannot place fire duties above safe passage. The same is true for rescue personnel.

The first overall observation and analysis on arrival at the scene is critical. This starts as you approach the scene. Slow the approach and observe the overall picture. Begin to categorize your priorities. It may be quickly obvious at a distance that additional help is required, perhaps from the fire or police departments. Call for this assistance before getting tied up in actually managing the investigation. The entire evaluation time to this point may be a minute, if done properly. The delay is minimal, and critical first actions are ensured. Once involved in activities at the scene of the mishap, it may be some time before the investigator can find time to seek additional services. There may be a need for immediate crowd control or to block off the area. It may be too much for one person to handle. Individuals at the scene may be pressed into service as volunteers to go for help, block off the area, or control onlookers. These helpers will need specific instructions on what to do.

If injured are still at the scene, volunteers might help. Often a medically trained person may only be waiting to be asked to assist. Make certain that the person is capable of any minor medical task assigned. He may carry a card testifying to first aid or resuscitation training. A safety person or investigator should be able to evaluate his ability after a few seconds of observing the person at work.

Here are the priorities for the investigator arriving at a mishap scene:

1. Arrive safely.
2. Observe the overall scene on arrival and evaluate the situation.
3. Care for the injured.
4. Protect others from injury and protect property from further damage.

In no case should you jeopardize yourself or others when professional help is at hand such as police, firefighters, medical personnel, and rescue personnel. While the investigation is important, it should never hinder the four activities listed above. Do not approach the scene until others have finished their tasks and it is safe to approach. Only then should the investigation start.

Another set of priorities may now be established. Their order is not concrete, and individual judgment is required. They may be carried out simultaneously.

1. Preserve evidence.
2. Protect the mishap site.
3. Secure the evidence.
4. Keep the boss informed.

Preservation of Evidence

When a mishap occurs it may seem to be the social event of the season, with people running in all directions, forgetting their work and abandoning jobs and cars. The confusion is compounded by fire and rescue vehicles rushing to the scene with appropriate lights and sounds. Add to this the people who have a bona fide interest and need to be present. The scene can be chaotic. Perhaps traffic is backed up and at the same time the media are pressing their way to the forefront with equipment. Few things are more disruptive to the conduct of business than a mishap, particularly a spectacular one. Preserving evidence and controlling activities under these conditions may seem hopeless. Yet this is a common situation, and the need to preserve and control is nearly always present.

Evidence gets washed away and trampled on by cleanup crews. Important debris vanishes into trashcans or is taken away as souvenirs. Skid marks and wheel tracks are erased. Witnesses get lost in the crowd or leave for work or home. Items and materials, that, if left in place, would provide valuable information, are moved and their value to you is diminished or lost.

More mishaps may occur as a result of the event taking attention from the primary task. If no one is in charge of the disorder, no attempt is made to bring order out of the chaos. Interest in the event gradually dies off as the injured are cared for, as debris is cleared away, as eyewitnesses move on, and as business returns to normal. The investigator may inherit only a few skid marks and an empty site. Early action is needed to deal with the event and find the causes.

Let us return to square one, the mishap plan, put together before the mishap ever occurred. A good plan will bring order, provide for the quick arrival of those who must save lives and protect people and property, and help preserve evidence for investigation. The preceding paragraphs describe an event involving many people, probably in a public area or crowded factory. The same types of things happen on a reduced scale in someone's kitchen, garage, or warehouse. The evidence vanishes unless someone takes charge and has a plan to preserve it. It does not matter if the area is remote. More backpackers, natives, sightseers, and scavengers will appear at a remote mishap site than have been seen in that area in years. Quick arrival at a mishap scene by the involved persons is desired, and they must be able to secure and protect the site. Since the investigator is seldom as quickly on the site as personnel responsible for first aid, rescue efforts, or firefighting, these people should be given instruction in crowd control and in disturbing the site as little as possible so that evidence can be preserved. This mission is secondary to saving lives and preserving and preventing loss of resources, but coordination with these agencies does much to preserve the scene. An hour of instruction to company foremen and supervisors on securing the mishap site and site control is worth several hours of instruction on mishap investigation. Site control should be part of the mishap plan and part of the instruction in getting ready for a mishap.

A critical question for the investigator is: How much time do I have? That is, How long can I spend at the site searching for the causes of the mishap? The time available for the investigation depends on many things. Public property must be returned to public use. Traffic flow must be resumed, and customers need access to a business. Machines must return to operation to continue production. There may be only time for a few good photographs and a sketch or two. Perhaps evidence will be gathered for later sorting and study. A company's policy may allow little investigation time except in cases thought to have special consequences, such as public interest or court involvement.

The first stakeholders to arrive should protect the mishap scene by keeping our unauthorized persons and keeping traffic to a minimum. Use bystanders to help with site control. It is not a bad idea to use witnesses. Keep them separated for later interviewing. It puts them under your control and keeps them available.

Who belongs to the mishap site? Only the investigator(s) belong there after fire, rescue, and security personnel are finished with their work. The investigator may need special help such as a photographer, safety director, recorder, and the like. Hold the onlookers to a minimum, and limit the participants to those needed for the job. Some evidence may not actually be at the site, but it is the mishap site evidence that is the most transitory and tends to disappear first.

A last thought on site protection and saving evidence: Due to the nature of the mishap debris or process, items of value may be involved. If so, guard the site to deter those who would steal or alter evidence.

Gathering the Evidence

The details of gathering evidence are covered in later chapters; however, those processes are greatly speeded by plans made before the mishap occurs.

Much of the evidence at a mishap scene is of a fleeting nature and is thus referred to as short-lived or transitory evidence. The investigator must act quickly to gather it. Rough sketches of the mishap layout, photographs soon after arrival, and a sharp eye will do much for later reconstruction of the scene.

Evidence such as skid marks, fire damage, scuff marks, some liquids, and scattered debris can disappear in a few minutes. Premishap plans should include gathering samples of fuel, hydraulic fluid, lubricating oils, broken pieces of glass, and other possible evidence. Small pill boxes, envelopes, small glass containers, plastic envelopes, and similar containers, will aid this effort. Each sample should be clearly labeled as it is gathered, and the location from which it was taken carefully marked and noted.

Pressures to return the scene to normal call for sweeping up, hosing down, and picking up debris. Skid marks, fire damage, scuff marks, and liquid pools can be noted by quick sketches or photographs. Rough distance measurements should be taken of the various elements involved. As much of this as seems critical should be handled at once so that the evidence will be properly tagged and packaged for later examination.

Where liquid pools will soak into the ground, they should be sampled quickly. This means including the soil if that is the only way to get a sample. The extraneous material should be identified as far as possible to assist in the analysis. Tight control of samples is essential even when they are sent off to a laboratory for analysis. It is critical that the analyst know as much as possible about the scene and situation that can aid in making an evaluation.

Do not overlook the need to gather human evidence such as blood and breath samples. Special care must be taken where the individuals may have been influenced by some drug. This too is short-lived evidence requiring quick action. Samples of human hair or clothes worn by individuals require special care to ensure their proper preservation and analysis.

Evidence should not be cleaned by the investigator before being microscopically examined. Wiping off suspect metal surfaces or hosing off dirt or soot can greatly reduce the chances of a good analysis. Possibly the only exception to the cleaning process is where human excretment or dangerous chemicals are found. A better way is to handle the affected items carefully with gloves, package them tightly, and clearly identify the situation to the examining laboratory. When a mishap occurs at night, every effort should be made to retain the scene as is until careful examination can be made in good light. If this is not feasible, portable lighting can sometimes be rented from organizations prepared to handle emergencies.

Documents are often important evidence, whether found at the scene of a mishap or elsewhere. The investigator may find some of these to be of a fleeting nature also. Items such as trip tickets, maintenance records, and personnel training records tend to get lost, destroyed, or transferred, even though the person or machine is not expected to return to service. The investigator should have first claim to such evidence even though litigation or personnel action may be expected later.

All evidence should be tagged and kept separately. There are a few guidelines for handling certain kinds of evidence. Paint, a common type of evidence, is involved in most physical contact cases. It is impossible for a transfer of molecules not to take place[2] when materials rub together. A minute piece of paint is bound to be transferred when a painted object makes contact with another object, although it may not be visible to the naked eye. A microscopic examination may be needed, particularly if a piece of wood or clothing is involved. To match up paint samples through analysis, a chip of paint a quarter-inch square is desired. Even less will allow a color match-up or rubbing analysis. The paint must be scraped carefully into an envelope and sealed in, and identification must be given on the container or sealing tape.

When soil analysis is desired, as little as a teaspoonful may be enough. The soil should be dried before being placed in a container to avoid complications of mold growth. If the soil is to be matched with that from another location, several samples to cover the area should be submitted. If the soil is attached to an object that can be packaged, then the whole thing should be submitted.

When an individual has been struck by a vehicle or machinery, we can expect pieces of his clothing or at least imprints of the material to be on the machinery. Once the suspect vehicle and the location of the imprint are known, microscopic examination can confirm or deny the evidence. The evidence on both the equipment and the clothing worn should be sent for analysis.

Glass presents unique opportunities for analysis. Broken parts can often be matched or otherwise identified as coming from the same equipment. Under some circumstances pieces of glass can be identified as coming from a particular vehicle or instrument. Police laboratories are particularly skilled in this type of analysis.

To sum up, evidence is often short-lived, and advance plans must be made to aid in gathering evidence. Nearly any evidence, if properly collected, can be analyzed for mishap involvement. It should also be noted that knowing that something was not involved in a mishap may be just as important as knowing that something else did play a role.

Data Recorders

Data recorders are covered at this point because it is important to quickly capture and protect their contents after a mishap. Data recorders are increasingly being used in investigations. Reading pressure records after a boiler explosion or a truck log after a highway mishap is only a start. The value of these and more advanced recording modes can be appreciated by looking at the aerospace industry. Data recorders are common accessories in aircraft, their major function being to provide data in the event of a mishap. While some airborne flight recorders provide information on as few as five activities, newer models can furnish information on a hundred parameters. Space vehicles commonly record hundreds and even thousands of activities, returning data to earth through telemetry. Typically such computers furnish data on operating parameters, operator input, systems status, engine performance indications, and control and switch positions.[3]

Analyzing the recorder content requires a thorough knowledge of the limitations and potential for error. The potential sources for error are many. Sometimes they are apparent while at other times they may be subtle and barely discernible.[4]

The use of a voice recorder in combination with the latest data recorder can help solve even the knottiest investigation problems. We have seen this in the space shuttle *Challenger* catastrophe. Its value had actually been proved decades before in an unmanned space flight of the *Apollo 6,* which had trouble in space and was never seen again.[5] Since there was no hardware evidence, the loss was analyzed by telemetered data and simulations. This telemetry trackdown of the 1,500,000 systems in *Apollo 6* resulted in tracing the problem to an engineer who had placed a wrong tube in a fuel system.

The increasing use of the data recorder in all transport and many industrial modes has brought its mishap investigation value to the forefront. For example, a nuclear facility mishap has been investigated in depth by data readouts, with all the causes thoroughly determined and preventive actions taken. The hardware still could not be inspected a year after the mishap. When crew members or operators are killed, their testimony is lost. If they survive, it may be unreliable or limited.

In the future we should expect to see greater use of the technique of combining voice and status recorders with simulation. Just coming into use, this technique connects recorder evidence with a device such as a flight or missile simulator to give a graphic presentation of what happened. Sometimes the simulation will not be needed since the data provided by the recorder will clearly show a situation.

Data recording and telemetry equipment is usually very costly, but it is a valued and useful investigative tool even though experts may be required to interpret the data. Recorder data should not be used in isolation. The investigation must provide factual data against which the recorder data can be measured. Without this corroborating information the recorder data will not be valid.[4]

A related development seeing successful field use deals with retrieving information from computer systems. Formerly, when a computer system was shut off by the operator or by a mishap, the data were lost. Now some systems retain data in their memory even when electric power is shut off. Such memories are termed "nonvolatile." Nonvolatile memories are found in some microcircuit chips or in larger modules containing serial core memory networks. The smaller chips have a good record for surviving serious mishaps. This subject is discussed in more detail in Chapter 7.[6,7]

SUMMARY

A most effective method to ensure a good investigation is to carefully plan for the event while knowing it might never happen and certainly will not happen as expected. At a minimum, the plan should provide for notifying the right people when there has been a mishap, mounting a rescue and resource-saving effort, and assuring a timely investigation. The plan can be short or detailed. The leaner plan tends to be more efficient and easier to update. Where the plan is purposely kept short to minimize problems in activating it, supplementary documents may provide details on what to do after critical first actions at the mishap site. Examples of detailed premishap plans may be found in many industries, at most military bases, and, for transport mishaps, from the Department of Transportation's Transportation Safety Institute at Oklahoma City.

QUESTIONS FOR REVIEW AND DISCUSSION

1 In your own organization, how many mishaps are investigated each year on average, and by whom?

2 Does your organization have the equivalent of a premishap plan? How many pages long is it?

3 How is the fire department notified of a mishap in your organization?

4 In your organization, is a mishap counted against the supervisor in whose department it happens? Is he in fact held accountable in some way for mishaps? How?

5 Has anyone in your organization been specifically trained in mishap investigation? Who? Where did they get their training?

QUESTIONS FOR FURTHER STUDY

1. Compare the accident rates in your organization with the national average for similar operations.
2. Analyze the mishap plan for your organization against the requirements outlined in this chapter. How does it compare?
3. If the supervisor in your organization must investigate a mishap, how does he get pictures of the mishap scene? Describe the exact procedure he follows.
4. Write a paragraph detailing how the public relations aspects of a mishap would be handled in your organization. Tell who handles it and what the plan is.
5. How far and where would you have to go to secure training in mishap investigation? When and where would the next class start?
6. Cite and describe an actual case where the nonvolatile memory of a computer served to solve a mishap.

REFERENCES

1. Kuhlman, Raymond L., *Professional Accident Investigation,* Institute Press, Loganville, GA, 1977, pp. 56–57.
2. Basham, Donald J., *Traffic Accident Management,* Charles C. Thomas, Springfield, IL, 1979, pp. 22–31.
3. Kaye, Michael J., "Flight Data Recorders and Simulation," *Aerospace Safety,* February 1986, pp. 10–15.
4. Grossi, Dennis R., "Analysis of Flight Data Recorder Information," *ISASI Forum,* October 1986, p. 35.
5. Lederer, Jerome, personal correspondence on this and other subjects, from Laguna Hills, California, Dec. 19–22, 1979.
6. Hazlett, Jack, "Recording Systems That Remember," *Air Force Safety Journal,* July 1986, pp. 7–13.
7. Conversations with Major Chet Cartwright at Norton AFB, California, September 1986.

CHAPTER THREE

Getting Underway with the Investigation

The first few minutes at the mishap scene are critical for the investigator. With a good mishap plan prepared in advance, you can move quickly and confidently into the investigation. The first actions suggested in Chapter 2 are good guidance for any investigation. They deserve repeating as we get ready for action.

1 **Arrive Safely at the Scene.** Do not go off half-cocked in the urgency of the event. Know exactly where you are going and what you will need when you arrive. Take a few minutes to methodically gather the investigation kit, notify the proper people, and ensure that firefighting and rescue efforts have started if they are needed. Arrive at the scene prepared to go to work.
2 **Observe and Analyze the Situation.** As you approach the scene, look for danger signs, obstacles to investigation, action needed to protect lives and property, and evidence that must be protected. In other words, get the big picture and see what must be done before starting to investigate. A trained observer can do this in a minute.
3 **Care for the Injured and Protect Others.** Help the injured, and keep others from harm and from interfering with the investigation process.
4 **Protect Property.** Do what is necessary to protect property from more damage.

TAKE CHARGE

Take charge if it is your job to take charge, but wait for professionals (firefighters, police) to complete their work. Do not get involved until the lifesaving effort is complete and the site is safe to approach, unless your help is essential and you can make a definite contribution.

While the investigation has a high priority, it may be proper for fire or security people to have overall responsibility. If this is so, seek their cooperation in carrying out the investigation. The nature of the mishap may require a higher authority (such as a state agency) to investigate and thus have a priority. Even when the investigator is in charge at the scene, cooperation is a powerful tool for success.

When the mishap investigator cannot take charge, which is often the case in mishaps with potential of danger to the public, cooperation and coordination are essential. Mishaps that endanger the public tend to involve outside agencies, none of which have full on-the-scene responsibility. If radiation, chemical spills, explosion, or similar dangers are possible, various assistance groups may be stakeholders. Even with these governmental groups, no one in particular may be in charge. As mishap investigator, you should make every effort to establish and clearly identify the on-scene commander and command post. Even if you cannot assume this role yourself, you should be ready to establish and support an on-scene activity to coordinate action. When postmishap activities involves several local, county, state, and even federal agencies, they often work in each other's area of concern and cannot legally take charge. Their efforts must be coordinated. Prior mishap planning will at least help the situation.

People should know by announcement or sign who is in charge at a mishap—if there is one specific person. If not, they should at least know where the control and coordination center is. Let them know if you are in charge of the investigation. This allows a center of action, a contact point for information, and helps you find out what is going on. Be positive, and take responsibility for the details, such as dealing with the news media, standing firm with persons anxious to restore the site to normal, and coping with the pressures of higher management. The sooner these details are attended to, the sooner you can get on with the investigation.

If you are in charge of a team effort, your job is easier in some ways and more difficult in others. It is easier in that you have help and the benefit of others' observations and opinions. It is harder because a team effort must be managed. The investigator in charge must do two main things: keep everyone working efficiently toward clear objectives, and assure good information exchange among all team members. The team should meet at the start of the day, review the plan of the day, and assign duties such as interviewing witnesses and forwarding evidence to the laboratory. At the end of the working day the group meets again, this time to trade information, discuss problems, and seek the best approach to the tasks that remain. This continues until the job is finished. As the work goes ahead, some specialists will finish their tasks and should be released to their regular jobs. As each member finishes, he or she should make a complete report on his or her work and even prepare part of the formal report. It is harder to get this information once a team member has left the scene and returned to work. Keep up the momentum of the investigation. It requires good management skills to keep everyone moving forward to completion as the task becomes old and frustrating. If the team is away from home, these problems often multiply.

PROCEED QUICKLY WITH CAUTION

The essence of gathering transitory evidence and returning to normal operation depends on quick action by the investigator. Safety cannot be compromised in the process.[1] It does not matter what speed is used to get to the scene if the investigator is delayed or injured as a result of that speed. While timely action is desired, remember that the investigator is not part of the rescue operation. The common-sense need is to arrive safely, in a timely manner, and equipped to investigate thoroughly. Timely common sense is better than hurried unpreparedness. The military service has a saying to the effect that "if you cannot get there immediately, calm down and be prepared to do a thorough job when you do arrive." Where speed is essential and the mishaps tend to be catastrophic, this is sound advice.

The best approach ensures that the mishap plan provides for rescue, preservation of property and evidence, and securing of the mishap site. People likely to be involved when a mishap occurs should be kept up to date on the plan. This makes the investigator's job much easier.

WITNESSES

We usually think of a witness as one who has seen the mishap or was near enough to arrive soon afterwards and furnish helpful information. A witness can be anyone or anything that provides knowledge about the mishap. In that sense a witness may not have seen or heard the mishap, not even know it happened, and yet have valuable information.

Since a witness need not be human, consider one technique that classifies witnesses into four types. Known as the four P's, the types are people, parts, position, and paper.

1. **People.** They need not be eyewitnesses or participants. They can be maintenance persons, doctors, supervisors, engineers, designers, friends, relatives, or anyone whose information can aid the investigation process.
2. **Parts.** This refers to failed machinery, communication system failures, inadequate support equipment, improper fuels and lubricants, or debris at the mishap site.
3. **Position.** This concerns the mishap location and involves the weather, roadway, operating conditions, and location, trable direction, wreckage resting position, and the like.
4. **Paper.** Paper performs as a witness through records, publication, tapes, directives, drawings, reports, and recordings. Nowadays we might also include computer software.

In this chapter the emphasis is on the human being as a witness, but the investigator must be aware of the need to coordinate a mass and variety of witness information from different sources.

Gathering eyewitness facts can be time-consuming and expensive, and the facts gained can be contradictory. It is said that in the average case we can expect half of our information to come from human witnesses and to be only 50 percent effective. The effectiveness of the other 50 percent depends on how well the investigator evaluates, analyzes, and ties information together. Since eyewitnesses provide over half of our information, their value is great. Realize that witnesses seldon see all that has happened. The events leading to a mishap are often routine and no cause for concern or they are so unexpected and chaotic that little of value is noticed.[2]

Promptness

Witnesses should be interviewed as soon as practical to ensure the integrity of their information. One reason for the high cost of mishap investigation, which may be more than the cost of the mishap, is poor handling of witnesses. Getting to the witness quickly helps get better information.

Study of the time factor in interviewing gives us some points to consider. Witnesses are strongly influenced by each other and by the news media. When given time to talk to or listen to others who have seen the event they may change their story. They can listen to a TV news account and alter their story to support the media account. A witness may not understand what he has seen and, innocently, to explain things in his own mind, will change his story. This can happen unconsciously and without any intent to deceive. Several hours later or the next day he may have a different story to tell and believe it himself. Witnesses can also forget. A witness may not remember some events, particularly if they do not confirm what he has come to believe he saw or heard. Minor details are forgotten and are therefore lost to the investigator. Some witnesses are hostile and, given time, may change their story to hurt the company or protect a friend. Some witnesses become indignant when they have something to say and a long period elapses before they are approached. They may then claim to have seen nothing. Prompt interviewing is the best solution.

The longer a witness sits on his statement, the more likley he is to modify it. Information given soon after a mishap will have more details. After a couple of days the information is much more general and has only a few details. After a week, only the most vivid details are retained. Positiveness also declines with the passage of time.

Sometimes the witness tells friends about the exciting event. The more he tells it, the more settled he becomes on what he thinks he saw. The changes in the story are not deliberate attempts to tamper with the truth, but as the story is repeated the witness's ideas becomes more concrete.

The best advice is to get a statement quickly and, if time is short, fill in details at a later interview. You can keep witnesses apart by giving them tasks at separate site locations. If you do nothing else, have them write up what they saw and/or draw a sketch of the scene.

Reliability, Probability, and Uncertainty

The witness sometimes has a built-in problem deciding what he saw. Consider this research showing that a witness is influenced by several things. A cab was involved in a hit-and-run accident with a bus. Two cab companies serve the city, the Yellow, which operate 85 percent of the cabs, and the Red, which operate the remaining 15 percent. A witness identifies the hit-and-run cab as Red. When the court tests the reliability of the witness under circumstances similar to those on the night of the accident, he correctly identifies the color of the cab 80 percent of the time and misidentifies it the other 20 percent. What's the probability that the cab involved in the accident was Red as the witness stated?

Most will state that the probability is 80 percent Red, but in fact the cab was likely Yellow. Here is why. Imagine that the witness sees 100 hit-and-run accidents instead of just one. By the laws of probability, about 85 of them will involve Yellow cabs and 15 Red. Of the 85 Yellow cabs the witness will misidentify 20 percent, or 17 cabs, as Red. Of the 15 Red cabs he'll correctly identify only 80 percent, or 12. Thus the 29 times the witness sees a Red cab, he's wrong 17 times—an error rate of nearly 60 percent. The base rate—the preponderance of Yellow—makes the odds 60 to 40 that he will misidentify a Yellow cab instead of a Red one.[3]

Or suppose you are taken into a darkened room and shown a circle with no distance cues. Your are asked how big it is. You don't know whether it's a small circle very close of a large circle far away. But people always have a firm feeling about a circle's size even when they know they're wrong. This is a classic example of how the human mind suppresses uncertainty.

Locating Witnesses

Early arrival at the mishap scene provides the best chance to locate eyewitnesses. An investigator pressed for time because of other duties can at least get names, addresses, and telephone numbers. Many witnesses will be transients, and there will not be a second chance to contact them. Interview these people first.

The files are full of cases at first it seemed there were no witnesses to a mishap, but a thorough search later turned up good eyewitnesses. In one case where 117 persons were near a mishap, 116 of them denied seeing anything. Only when the 117th witness was interviewed did it surface that he had seen the entire event. In another case a witness stepped forward 10 years after an event. For that entire period he thought his evidence had no value. Such cases are common and should alert us that there are few mishaps without a witness of some kind.

Persons actually involved in the mishap are the investigator's first priority. Next comes the eyewitnesses who, incidentally, can oftern name other witnesses. For both of these, be sure to determine where they can be reached later.

It may be necessary to canvas an area for witnesses. Civil workers such as police, rescue crews, or firefighters may have been in the area and seen the event. Appeals for witnesses to come forward may be made over radio and television or through newspaper advertisements. Caution is needed, since some who come forward will simply be seeking publicity and attention. Others will have personal reasons for not wanting to come forward.

Qualifying the Witness

The value of a witness's statement depends partly on his credibility. Giving value to a witness's statement is done through qualifying him, that is, finding if the witness has seen or said is of value. In a court this is done by examining credentials and experience. Those fully qualified to make judgments are called expert witnesses. Typically, a witness may not be an expert in the courtroom sense and yet be fully able to recount what he has seen without distortion. It helps to know a person's background, location at the time of the mishap, and other aspects relating to the mishap environment.

We would expect a pilot to be credible with information about an aircraft mishap. If a lathe operator is hurt, we would expect the operator at the next machine to be a good witness. A highway patrol officer who sees a delivery truck have a mishap should be a good witness. However, because of their expertise they may expect certain things to happen. When they don't, the witness has trouble reconciling what was seen with what was expected. A well-qualified witness may have a high level of confidence in his statement and seem to have all the facts. However, being confident does not ensure the accuracy or validity of the information.

Due to closeness to the event or to some involved person, otherwise well-qualified witnesses may not view the event objectively. For example, a mother might not be able to describe an event in which her child was hit be a car.[3]

All things being equal, when two witnesses see the same thing at the same time, we would expect the one knowing the most about the situation or equipment to give the better statement. In any case, a witness account must be verified and substantiated. The statement should be considered in the light of the witness's experience, relationship to work, personal involvement, and agreement with other witnesses.

Interviewing the Witness

The question is whether to interrogate or interview. Should you approach the witness in the formal, oftern highly structured approach of an interrogator, or the open, receptive, "let's hear what you can tell us" approach of the interviewer? There are advantages to both.

The interrogation is more likely to reflect an adversary meeting as in a police-suspect relationship. It often uses a highly structured, sometimes incriminating set of questions. It may in involve more than one interrogator to make certain nothing is overlooked.

The interview is usually less formal, meant to put the witness at ease and assure him that he is among friends, people who need his help. The interview, being less structured, tends to be more time-consuming. The relaxed atmosphere with one investigator should make the witness more willing to talk and answer questions that might involve him in the mishap. The intent of interviewing—to prevent more mishaps—is made clear to the witness and implies that something good will be done with the information.

Interrogation and interviewing are sometimes called structured and unstructured questioning. "Structured" does not necessarily imply an adversarial situation, only that the investigator presents a list of questions. It is hard to say which technique is best. Research tells us that no definite conclusions can be reached either way. The success of either approach depends on the skill of the investigator. The skilled investigator will likely use a combination of both, depending on the situation, relaxing as appropriate and pressing as needed to get information.

Most investigators with prevention as their goal prefer a low-key interview approach. This elicits the most information, but with the most errors. The witness is placed at ease. He is deliberately not pushed, proceeds at his own pace, and at first receives only general guidance. The witness is made comfortable with good seating and small talk in a quiet, nonthreatening environment. He is encouraged to tell what he has seen, entirely in his own words and without interruption. The need for his statements is made clear, and he is encouraged to tell all. The investigator nods, listens closely, and encourages the witness. Note-taking or recording is kept unobtrusive. The investigator should be a "good guy" seeking help. The adversary role should be resisted unless a change in approach is needed. Try to confine the witness to observations and discourage conclusions. Conclusions are your job.

The witness is seldom required to make a statement on a mishap (see Chapter 18). His actions are voluntary. He has information, and you want it. If he does not like the way he is being approached he may resist giving a statement and even walk away. Witnesses who speak only a foreign language may be especially sensitive to the way they are being treated. Give them the same consideration through a translator and provide for their privacy during the interview.

Types of Witnesses[4]

Some witnesses are shy and retiring. They should be placed at ease, made to feel that they are of great service, and not pressed too hard. When the investigator has the witness's confidence, things usually go well.

The hostile witness may not come forward and may refuse to cooperate when found. He may not respond to friendly overtures. Since he cannot be forced to cooperate, it calls for much skill to secure information. The hostile witness may have something against the company or people involved and may deliberately mislead or distort information. His statements require careful evaluation. He may be seeking to get someone into trouble.

Nearly every mishap investigation has the person who wants to be involved for egotistical reasons, to become part of the "scene" or because he thinks he has superior knowledge about the subject. Some parties seek this type of involvement even though they are far removed from the scene and have little knowledge of the event. A careful review of the facts will help you deal with this witness. Control him by having him place events in a definite sequence, and later have him repeat the sequence.

Sometimes the witness is impaired by alcohol. To put it bluntly, if he is drunk, forget him; he is not worth the trouble. Wait until later, and see if he can recall anything when sober.

If the witness is sick or injured, take care not to endanger him in any way. Consider his impaired condition when evaluating his statements. If he is an operator, his statement may be the most important of all, but his condition must not be worsened. If you must question this witness, plan carefully and limit the questions. Good judgment means weighing the advisability of taking an early statement against worsening his condition. An early hospital interview is better than an ambulance-side exchange. Follow the advice of medical personnel.

Children are both good and bad witnesses. At 8 or 9 years of age, they have a few preconceived notions and pretty much tell it as they see it. Younger children have many fantasies and embellish what they think they have seen. Always have a parent's permission to talk to youngsters.

A person whose livelihood depends on the weather should be a good witness on the subject. Farmers, for example, are sensitive to the weather. Their livelihood depends on it, and it controls their daily schedule.

If possible, talk to witnesses where they saw the event so that light, angles, and distances can be confirmed. This also gives the witness clues on what he saw or heard. Witnesses who were in different locations will have different details of the mishap.

Witness Manipulation[3]

It is easy to manipulate people's memory by asking leading questions or by exposing them to new or leading information. While an attorney may have good reason to manipulate someone's memory and witnesses' statements, this book recommends avoiding it in the search for factual information.

Leading questions can subtly manipulate memory. Asking "What did you see?" tends to get more accurate information than "Did you see a gun?" One researcher cites a case in which subjects who were asked to guess how "tall" a basketball player was estimated an average of 71 inches. When asked how "short" he was, a similar group estimated 69 inches. Even the use of "a" or "the" can affect the answers. When asked "Did you see the broken headlight?" people were more likely to respond yes, they had, even though there was no broken headlight, than those asked "Did you see a broken headlight?" "Some" or "any" can make a difference. A questioner uses "some" when he expects a positive response and "any" when he wants a negative response.

Hypnosis is sometimes used to manipulate witnesses into recalling events thought to be in the permanent memory. However, a growing controversy over the nature of long-term memory has led to rulings in several states that hypnosis cannot be used in criminal cases. On the other hand, there are many cases where a witness has been traumatized by an event and subconsciously pushed details on it out of conscious awareness. Then hypnosis has broken through and allowed the details to be extracted. The controversy rages, based mainly on the idea that under hypnosis a person is highly susceptible to suggestion and will try hard to please the hypnotist. What was not seen by the witness cannot be remembered. Neither the hypnotist nor the person being hypnotized can distinguish between true memories and pseudomemories in the recall.

Techniques of Interviewing

It would be impossible to discuss all the interviewing techniques that can be used in all possible situations. Listed below are some broad guidelines to be used with care.

1. Before starting an interview, have questions ready to ask after the witness tells his entire story. Interview in a quiet, neutral, non-threatening location.
2. Make certain the witness knows the purpose of the interview. For most of us the purpose is to prevent more mishaps and keep others from being injured. He may react favorably if he is simply told you are trying to find out more about the mishap. These reasons should calm his fear about getting someone in trouble. Don't mislead the witness about the interview purpose. Inform him fully, and do not spring any surprises about the nature of the interview.
3. Get essential information on the record: name, address, where the witness can be reached, etc. Try to do this before you begin the interview.

4 Let the witness tell the story in his own terms. Have him tell of the event the way he saw it, in his own language and without interruption. You can start him width: "Tell me what you saw or did in the order it happened. Start with when you first knew about the mishap."

5 Talk to the witness in his terms, not yours. He may not know a lathe from a milking machine.

6 Use models or sketches to help the witness tell his story.

7 Use a cassette recorder if possible, but be unobtrusive. Explain why you are using the machine, and gain his approval. Make notes quietly and discreetly on questions you will wish to ask later, particularly on those things that may corroborate his statements or those of others.

8 If the witness does not know where to start his story, a good lead is, "What first called your attention to the event?" This should start him talking.

9 Avoid leading questions. If you press the witness to answer in a certain direction, since he is often eager to please he may respond in the way you indicate just to please you. If asked if the car is blue, the chances are excellent the witness will picture that and say yes, even if he does not recall the color. It is better to ask him to rephrase or repeat the statement so that you are not leading him.

10 Tact, courtesy, and diplomacy are good interview rules. Be neutral but interested, friendly, courteous, and businesslike. Be informal and easy, but direct. Approach the witness as an equal. Avoid unpleasant subjects such as taxes. politics, and racial issues. Urge the subject onward, but do not appear to hurry him. Do not begin serious interviewing until the witness is friendly. Seek cooperation. If he becomes hostile, discontinue and return to the soft sell.

11 At the end of the interview the witness can be asked about the fire and rescue response to help evaluate these services. Be certain to ask him if he knows anyone else who saw the event.

12 Time is important. The longer the details of an event are held, the more likely they will be modified by:

(*a*) Some details will be forgotten.

(*b*) Details not remembered right after an event will be filled in by imagined information to make it seem right.

(*c*) New information overheard or learned in discussion with others will change what was actually seen.

These are only a few interviewing guidelines, but they should assure success. We might note a few things about witnesses' statements. For instance, we seem to pay more attention to both similar statements and unique ones. Witness intelligence slightly enhances reliability but does not improve recall or the ability to observe. There is little difference between male or female reliability. Emotion and excitement produce distortion and exaggeration in verbal statements. A frightened witness or one affected by trauma might not recall even the most vivid events he witnessed. Memory of details both before and after a violent event is poorer than usual. It takes time to lay down a solid and complete memory trace, and the violence interferes with that process. The complex event, one that has many and sometimes simultaneous elements, is more difficult for the witness to take in fully. However, the longer a person is exposed to an event the better his memory will be. People almost always underestimate the length of time an incident lasted. Speed is also hard to estimate.

When all witness information has been gathered and analyzed it must be evaluated. Here the investigator must be alert to misinterpreting witness statements according to his own prejudices and opinions. For example, if he is strongly anti-alcohol he might tend to disregard the good statements of a witness who had a "couple of beers." A constantly open mind is needed. Guard against limiting the input of fresh information by your biases.[2]

Interactions

Witness statements may confirm, contradict, or interact with other evidence. They must be analyzed and evaluated for credibility. If there are only a few witnesses, this may be done without formally organizing the material. If many factors and statements are to be considered, charts may be used to organize the information. Consider a relatively simple mishap. A forklift collides with a worker pushing a handcart out of a blind intersection into a loading dock at night. The first information from the witnesses may be shown as summarized in Table 3.1. When combined with a sketch of the scene, this table is of value in organizing the information. It helps to show the interaction of witnesses and evaluate their credibility. It allows us to pinpoint differences of opinion and resolve conflicts. It also keeps the

Table 3.1 Location, Activity, and Observations of Those on Scene

Witness	Location	Activity	Observation
Smith	Truck at dock	Waiting for helper	Saw cart come down ramp; distracted
Jones	East end dock	Having a smoke	Saw forklift go past
Ortega	Warehouse	Waiting in cart	Saw cart speed down ramp and get hit
Rogers	Back of forklift	Forklift passenger	Distracted, waving at friend
Evans	West end of dock	Supervising	Saw cart in front of forklift
Englis	Entering warehouse	Pushing empty handcart	Saw other handcart gather speed and pass

Table 3.2 What Was Seen and Heard by Those on the Scene

	Heard					Saw					
Witness	Forklift Claxon	Truck Noise	Warning Shout	Machinery Noise	Loud Noise	Forklift Beacon	Light over Door	Forklift Headlamp	Handcart Coming	Collision	Brakes Applied
Smith		X			X				X		
Jones		X	X	X	X						
Oretga				X	X				X	X	
Rogers	X			X	X						X
Evans		X				X	X	X	X	X	X
Englis		X	X	X	X			X	X		

Table 3.3 Which Event Did the Witness See?

Witness	Event 1	Event 2	Event 3	Event 4	Event 5
Smith	X			X	
Jones	X	X	X	X	X
Ortega	X			X	X
Rogers	X			X	
Evans	X	X	X		
Englis	X			X	

investigator from subconsciously adopting or discarding a witness's statement because it agrees or conflicts with his own theories.

Where several types of information must be ordered, a somewhat more elaborate matrix may be appropriate, as in Table 3.2. In this table only one item can be verified with confidence—the loud noise. It would appear that we cannot confirm the sound of a claxon, a forklift beacon, or a light over the door, since only one person noted these prominent items.

Where events or facts are listed, a simple but effective matrix as shown by Table 3.3 (not related to the event above) can verify what events took place. There appears to be no question about events 1 and 4. There is not enough verification to give credibility to the other events.

When points are in conflict or more data are needed to pin down gaps in knowledge, follow-up interviews are in order. They should be made in the same spirit of inquiry as the first interview. Care must be taken to indicate that there is no thought of the witness lying or misleading the investigator. Indicate that there are some gaps in the information you have gathered that need a closer review. Make it plain you need the witness's help. This interview will be more structured than the first, since you know where data are lacking and what you need to find out. The witness may have only followed a natural tendency to exaggerate, elaborate, and distort when excited. Stay businesslike to keep him on track. It may be that he has merely put the events in the wrong sequence, and his statement needs to be made clear.

With experience an investigator develops interviewing skills. You learn to be a good juge of character and to evaluate a witness's state of mind from many signs. Although this is not as important as in the interrogation process, knowing how to judge a witness's state will advance the interview. The serious investigator should understand interviewing techniques and in particular the elements of body language. Many good books are available on the subject.

MISHAP PHOTOGRAPHY

This section tells how a photographer can help the investigator and how an investigator can best use photography in mishap investigation. It makes no attempt to instruct the reader in the intricacies of film and photographic equipment or in the proper techniques. Expert photography is a subject for study beyond the scope of this book. But we do discuss what you will need in the way of photographs and how to go about getting them.

The purpose of mishap photography is to verify and record damage and injuries. We want to provide a photo record of the scene, involved components, environmental considerations, and any evidence that will lead to finding all the causes of the mishap. We also want to educate those who review the report on the situation, conditions surrounding the event, and the operator's point of view, as well as illustrating major points that apply to the event.

Planning

Planning helps to make the best use of the time available for mishap photography. Photographs can capture short-lived evidence for later use. A common technique is to quickly take a few essential photos to support the investigation and then take specific ones later when pressures have been relaxed. Random picture taking is time-consuming and expensive. Knowing and planning the standard shots used in most investigations provides a better start than going ahead without any plan.

What if you arrive at the scene late? The investigator may not arrive for hours or even days after the mishap. Since most of the civilized world takes pictures, perhaps one or more people will have taken pictures soon after the mishap. A few questions in the area to the press and to routine transients may lead to some fine photos. Occasionally you will find that gem: a picture of the mishap taking place. This should be considered even where the investigator was on the scene early and has good pictures.

The camera can record an infinite amount of detail, often items an observer's eye will miss, and all in a few seconds. Proper planning of key shots early in the investigation will pay off.

There are some handicaps to photography. The poorly prepared investigator may go to the scene and shoot wildly in all directions hoping to capture some information, only to find he has none. After the wreckage is cleaned up we may find that, for some reason, our pictures did not come out. It is then too late to take pictures.

Only those photos that will be referenced in the body of the report and promise to be pertinent should be printed, but do take all the pictures that might be needed. A guideline is to take all the photos that could possibly help the investigation. The photographer should also be encouraged to take duplicate shots with different exposures or lighting. The economic secret is to take many pictures but only print the good ones. Stated another way: Overshoot and underprint. Film is cheap compared to the cost of printing and enlargement. That means that you either have to judge your pictures by looking at the negatives (most of us cannot do that) or work with proof sheets or contact prints.

If line supervisors or managers are to take mishap photos they should be taught the use of the equipment and the type of pictures needed. This is part of the planning process. The best camera for this situation is a "point and shoot," automatic everything, 35-mm camera. Where a company photographer will take the pictures, the mishap plan should provide for getting him to the scene quickly; this may include an after-hours arrangement.

There are some general guidelines for the first pictures to be taken at a mishap site[5] that apply to nearly all mishaps:

1 Take a few shots approaching the mishap scene (particularly a transport mishap) to catch the operator's point of view and transient evidence.

2 Photograph anything that is likely to be quickly displaced as soon as possible. This includes:
- (*a*) Medical evidence.
- (*b*) Personal effects and equipment.
- (*c*) Instrument readings and control positions.
- (*d*) Any evidence likely to be removed by weather, traffic, or cleanup crews, such as ground scars, heat evidence, and liquids.

3 At a fire in progress, take color pictures of flames and smoke. The investigator can analyze these for information on the type of burning material and temperatures. Pictures of onlookers may reveal arsonists.

4 An overhead view from a ladder, building, or (with widely dispersed debris) an aircraft will show the overall scene. Log the height of the shots. Try to get the shot as nearly vertical as possible and try to have something of known size in the scene. This will facilitate measurements.

5 General photos from all four sides are proper and are known as the "standard four." The distances from the subject should be the same and must be logged for the record.

6 Pictures made or orient the debris should be close enough to see the debris but far enough to show a relationship to other evidence. Consider this when showing ground scars and terrain features.

7 Show enough of all scenes to give a good orientation. In close-ups, include a familiar object such as a clipboard to show the scale involved. Including a ruler or tape measure is best for showing the scale. Take photos from the witness's viewpoint and, if possible, have the same lighting conditions.

8 Take close-up pictures of significant parts and/or fracture surfaces. Start with a wide-angle shot to show the part relationships, and progress to lab photos if needed.

The Photo Log

The busy investigator will have several problems after things have calmed down and he is analyzing the photographic evidence. Pictures that were meaningful at the time they were taken are difficult to sort out when the prints come back from the lab. This is even more of a problem when someone else has taken the photos. If you do not have a photo log and a method of identifying pictures, you are facing future embarrassment.[6] These are things you need to know about a picture:

1. When was this picture taken?
2. Who took it?
3. Under what conditions (film, light, lens) was it taken?
4. Where was it taken from?
5. What does it show?

Much of this information can be covered on a photo identification board that is photographed on the first shot of a roll of film. This identifies the roll and all negatives on it. It also identifies the date, photographer, subject, film, lens, etc. The subject of each shot is identified on the photo log and correlated with the negatives on that roll.

Expert or Amateur

Many investigators are experienced in photography, have good equipment, and take fine pictures. This provides an advantage in convenience, time, and money, and in getting the exact pictures desired.

There are several reasons, however, for turning to the professional photographer[7] for help, including lack of investigator capability. Specialized photography of equipment may be needed. There are few full-time mishap photographers, but some insurance companies and government agencies have persons who do this so often they can be thought of as professional mishap photographers. Some are on call to law firms, appraisers, and claims adjusters. They are not investigators, however, and do need guidance. Once told what is important, what is needed, and where to focus, they do excellent work. The investigator should provide guidance in duplicating scenes and bracketing evidence; there may not be a second chance to get the information. Using a professional photographer will not free the investigator, since he must guide the photographer. Several uses for the expert are given in the next section. The expert photographer can use many techniques, beyond those of the average picture taker, to duplicate the actual colors. One thing he won't want to face in court is saying "No, not exactly," when asked if his pictures do exactly duplicate the actual colors. It is common for the investigator to take pictures well within his capability and call on the expert for specialized technical work.

Specialized Applications

Color photography, more expensive than black and white, is often needed to show contrast and details. The film itself is not much more expensive. The real cost control lies in selective printing, not in taking fewer pictures. Color is essential for fire photography to show smoke color, flame intensity, and soot markings. Color film is used almost exclusively in medical photography and shows injuries well. Caution is needed to ensure that the true color is shown by the film and filters used. This is important where fluids and stains are identified by their color.

Macrophotographs are ultra-close-up pictures helpful in seeing debris fractures that need close study or to show the nature of scars and rubbing marks. Ultraviolet and infrared photography, when combined with certain filters, can show features not visible to the naked eye. Motion pictures can be used to record event reenactments. Video tape systems are commonly used at the site to show mishap details and the progress of the investigation, for later review and discussion of corrective measures.

X-ray photos of debris can reveal stress or breaking points. X rays allow one to look into sealed switches and valves to see operating conditions before disassembly for examination. Modern industrial x-ray equipment can do this well, and the common dental laboratory can do it in a pinch.

Thermal scanners and thermal video cameras operate in wavelengths beyond what the eye can see. They can be used to judge emitted heat, particularly soon after fires or explosions. This helps to pinpoint fire sources.

Microphotography might use micro lenses to enlarge fractured parts or other suspect areas, up to 8000×. The common technique these days uses a scanning electron microscope (SEM) with a special attached Polaroid camera.[6] An expert is needed to interpret microphotographs with large magnification. To the untrained eye, a polished surface, for example, may appear rough and unrecognizable in a microphotograph.

The expert photographic developer can, through exposure techniques, bring out details that would never be seen otherwise. He must know, however, what the investigator seeks.

Not within the scope of this book, but of utmost importance, is the use of lighting to bring out details of a mishap scene. The need to avoid reflection, to show highway marks in sun glare or at night, are but a couple of examples. It can be seen that good investigative photography calls for the amateur to be good or be able to call on professionals for help.

Equipment

The investigator will probably function best with the "all automatic" camera that automatically loads, automatically exposes, automatically focuses, and automatically advances the film. With a built-in flash it will automatically handle the lighting. Moving up a bit the photographer would probably go to a 35-mm through-the-lens camera, a strobe light, a couple of special filters, and maybe a wide-angle lens. The expert may have other preferences regarding lens, filters, lighting, film, and type of camera.[6]

Instant print cameras are undesirable because of limitations of print size, numbers of prints, enlargements, and close-ups. They might not be disqualified in court if they are the only pictures available, but an expert would probably have to be called upon to tell what the instant camera pictures show. One could expect to be closely questioned in court if this kind of photo were introduced on critical evidence.

Medical Photography

Medical photography is a highly specialized field and is most important when survival aspects of mishaps are investigated in detail. While it is good to have pictures of a body as it appeared in position following a mishap, detailed pictures of injuries can wait for the hospital or morgue. The reason for the on-site photography is that the unmoved body often tells a story about the cause of a death and becomes vital in insurance, injury prevention, and court actions. Most often it is best to leave medical photography to the experts.

Fire Photography

Fire photography is also a highly developed field, with experts in many fire departments. The rules for photographing fire start the same as for any accident. Take pictures approaching the scene, and take pictures from all sides.

Assuming the fire photographer will get shots of the fire in progress, some extra considerations are in order. Industrial fires, for example, may pose additional hazards of explosion, excessive heat, or poisonous gases. Using telephoto lenses for close-up views is preferable to exposing the equipment and photographer to hazardous conditions. Be careful when using a flash in areas where gas or explosive vapors are suspected, as the sparks from a poorly maintained flash could trigger an explosion. Early shots at a fire should record the use of breathing apparatus and protective clothing.[8]

While the fire is in progress a photographer should record the progress of the fire and attempts to control it. He should try to document the causes and effects of the fire, its point of origin, and the extent of initial involvement. Photographs may reveal that the fire started in more than one place. Documenting the presence of steam, color and density of smoke, and color and size of flames can help determine the nature of burning substances.

If the investigator has fire pictures he should use a professional fire investigator to interpret them. The expert can often tell such things as the origin of the fire, accelerants used, the nature of the burning material, and the fire intensity from just one good color photograph. After-the-fire pictures are valuable to fix fire sources, confirm arson, illustrate fire intensity, and provide other helpful information. For the investigator taking fire pictures some directions are suggested in the following paragraphs.

Take Color Pictures of Flames

Photographs of flames can indicate what material is burning, how the fire started, and the fire intensity. A yellowish to white flame indicates a hot flame of about 1500°C, while a reddish color indicates a heat of about 500°C. Running flames on water indicate petroleum products, and the flames would be red. (Appendix G provides more data on temperatures.) Color photography is of special value in nearly all phases of fire investigation.

Smoke Shows What Material Is Burning

Heavy black smoke usually means a petroleum product, rubber, or certain types of structures. Light white smoke signals combustibles such as wood, paper, or vegetation. An aura of brilliance around the base of the smoke indicates burning metal. The colors of smoke and flame are subject to variation, but the following may be helpful:

Combustible	Smoke Color	Color of Flames
Wood	Gray to brown	Yellow to red
Paper	Gray to brown	Yellow to red
Cloth	Gray to brown	Yellow to red
Gasoline	Black	Yellow to white
Naphtha	Brown to black	Straw to white
Benzine	White to gray	Yellow to white
Lubricating oil	Black	Yellow to white
Lacquer thinner	Brownish black	Yellow to red
Turpentine	Black to brown	Yellow to white
Acetone	Black	Blue
Cooking oil	Brown	Yellow
Kerosene	Black	Yellow

Spectators

It has been found that an arsonist often sticks around to watch his fire. An individual who shows up in a couple of pictures becomes suspect, so "shoot" the spectators.

Charred or Burned Areas

After the fire, photos should be made during successive stages of the clearing and search, with particular attention to burned and charred areas. Look for all types of incendiary devices and combustible materials, including candles, matchbooks, paper, cloth, and containers of liquids such as kerosene and paint thinners. Since the subjects and areas are often blackened and may reflect less light than their surroundings, special care must be taken to avoid underexposure.

Media Sources

The spectacular nature of fires makes it likely that good color coverage was made by TV stations or newspapers. They will usually cooperate in furnishing copies of their material.

Legal Considerations

An excellent legal discourse on the use of photographs for admissible court evidence is given by Donigan and Fisher.[9] The investigator is referred to that publication if the possibility of court use exists. Perhaps the only restrictions on taking pictures for investigative purposes involve a person's privacy or classified government activity. The investigator should have permission to go onto private property to take pictures. Where delicate matters of photography are involved, the best guidelines are common sense and good taste.

Photography Hints for the Investigator

The following checklist incorporates new material[10] with what we have covered so far in this chapter:

1. Photograph before touching anything. If waiting for a photographer, make walk-throughs and sketches.
2. Use color film for stains and marks on dark surfaces.
3. Infrared photography requires a special film and filter to go with regular equipment. Special protection must be given the film.
4. Ultraviolet photography will sometimes show fluorescent traces and alterations in documents.
5. Microphotographs are pictures taken through a microscope. Macrophotographs are simply larger-than-life pictures.
6. A spectograph is a photo device for inorganic substances that allows comparison with a standard.
7. When photographing imprints of materials on one another, take the picture from above and parallel to the imprint. Another photograph slightly to the side will show more detail but will be less accurate.
8. Almost anything of human origin should be photographed in color.
9. A picture showing what a person saw should be taken from eye level at the observer's position. Obstructions to vision may show up.

10 Photograph any equipment that may have played a role, such as steps, scaffolding, rigs, and support devices.
11 When bodies have been removed, outline the position with chalk or tape before taking pictures.
12 Panoramic photos should overlap about 50 percent before being fit together.
13 If you have a choice, use a background that provides contrast.
14 Have similar photos printed as a contact sheet, then make a choice.
15 Be careful of distortion from wide-angle and telephoto lenses.
16 Keep the camera level for easier orientation.
17 In close-up scenes use a ruler or other familiar object to show size. Use objects of known size in all photos to allow comparison.

SKETCHES, DIAGRAMS, AND CHARTS

A sketch is a drawing made at a mishap site, usually with data gathered in the field, that can later be transcribed into a more detailed and accurate form. A diagram is an arbitrary or stylized drawing showing sequences of action, process flow, and the like that bear on the investigation. A chart contains information, often statistics or measurements, helpful to the investigation; they are often extracts or copies of records such as speeds, instrument readings, and temperatures that might help the mishap report reviewer understand what has gone on. Charts may include functional information that clarifies certain points.

The same rules apply to diagrams, sketches, charts, and the like as apply to photographs. If they do not lead to a clearer understanding of the investigation and are not referenced in the report, they are not needed.

Sometimes drawings of the area are made to such a precise scale and with so much detail they are called maps. This is seen mostly in traffic mishap investigation. Excellent coverage on how to make maps of this kind is given in Baker.[7]

Sketches and diagrams are an important source of information. They tell the location of different "actors" in the mishap, and the movement and direction of people and equipment. So-called field sketches are done at the scene in a format suited for making them into maps and diagrams. Sketches are labeled with key directions, distances, and angles. This information may be listed in tabular form as a chart, with base points indicated in the sketch used as references for the measurements. Such measurements also locate key equipment and debris. In an outdoor mishap, features such as streams, hills, trees, and trails may be noted if they bear on the mishap.

A sketch or a chart may be more than we usually think of in mishaps. We might have, for example, plots of noise, light, temperature, or airflow.

Although it may seem basic, when making a field sketch with a pencil,

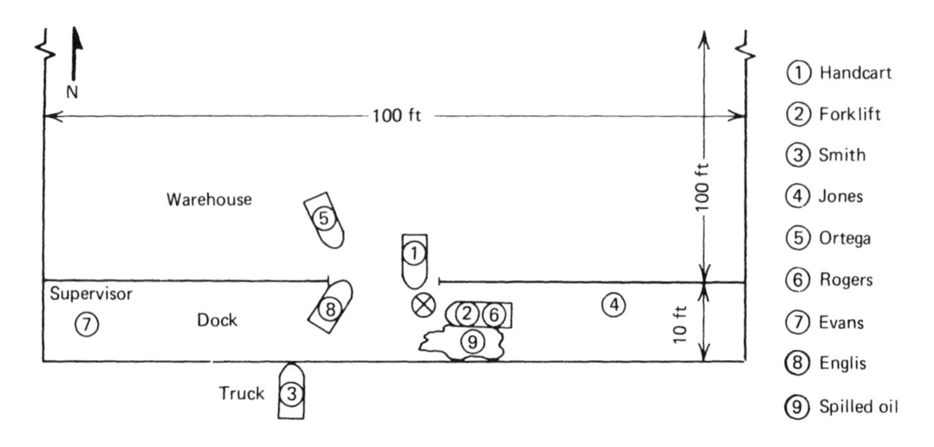

Figure 3.1 Transient information shown on a field sketch.

paper, tape, and angle-measuring device, precision is important. Key investigative issues or court action may hinge on the measurements. As a basis for investigative analysis, accuracy cannot be overstressed.

On arriving at the scene, transient data should be gathered first. Attention should also be given to short-lived evidence that will soon be gone. A simple field sketch with transient information is shown as Figure 3.1

In an open area where debris is scattered for some distance, a field sketch might look like Figure 3.2. The numbers in circles refer to items of either a transient or permanent nature and are explained in a legend beside the sketch.

Symbols used in mishap diagrams are seldom standard except in transport mishaps. A logical approach to symbols should be attempted. Baker[7] and the American National Standards Institute provide good guidance.

Drawings can best be prepared on grid or quad paper. Orient the paper to the major axis of the mishap and pick a scale that will include the area to be covered. This scale, since it is on a grid (e.g., one grid square = 10 ft), allows

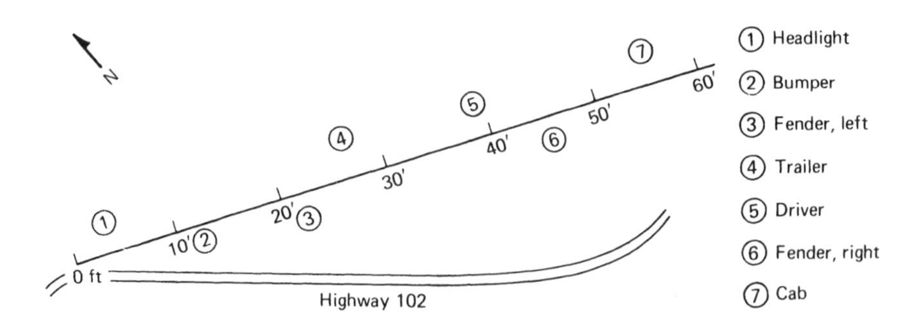

Figure 3.2 Field sketch showing debris scatter.

the investigator to make measurements and rapidly transfer them to the sketch by simply counting grid squares. It also permits you to make neater marginal notes and references. A centerline placed along the major mishap axis helps measure and locate individual items.

Some of the standards that have been developed for sketches and diagrams are the following:

1. Place the magnetic orientation in the upper left.
2. When convenient, the north–south orientation should parallel the left margin but not when it prevents clear understanding.
3. The lower left corner can have a square giving data such as weather, elevation, distance to the nearest landmark, etc.
4. The upper right should have information identifying the mishap, such as time, place, and investigator.
5. Diagrams drawn to scale should include direction of approach, first contacts, ground scars, and the like.
6. The diagrams should include location of major wreckage items, corpses, the wounded, key witnesses, etc.
7. Include key landmarks, terrain features, and distances between important points.
8. The location and distance from which key photos were taken should be given to serve as a basis for interpretation.

The above items apply to a plan view. A side view may also be needed. In a few cases an oblique or three-dimensional view is needed for clarification. Again, if there is no need for any of this, forget it. Placing this type of material in a report is a big temptation, and even more so if professional help has been used for the job.

GRADIENT MAPS

Environmental factors such as light, heat, and noise have long been recognized for their contributions to mishaps. Health-oriented safety professionals and others are familiar with mapping the contours of such effects through instrument and meter readings, but it remained for Kuhlman[4] to formally introduce gradient mapping as an investigative tool. As an example, suppose that a worker fails to heed a warning claxon and is struck by an electric cart. Investigating the mishap, you suspect noise as a reason for the worker failing to hear the warning, although others heard it. Using a sound-level meter you carefully take readings throughout the area. Connecting points of equal noise level, you end up with a contour map (Figure 3.3), not unlike a contour map of a land area, except that this one shows lines of equal noise level instead of lines of equal elevation. You note an unexpected

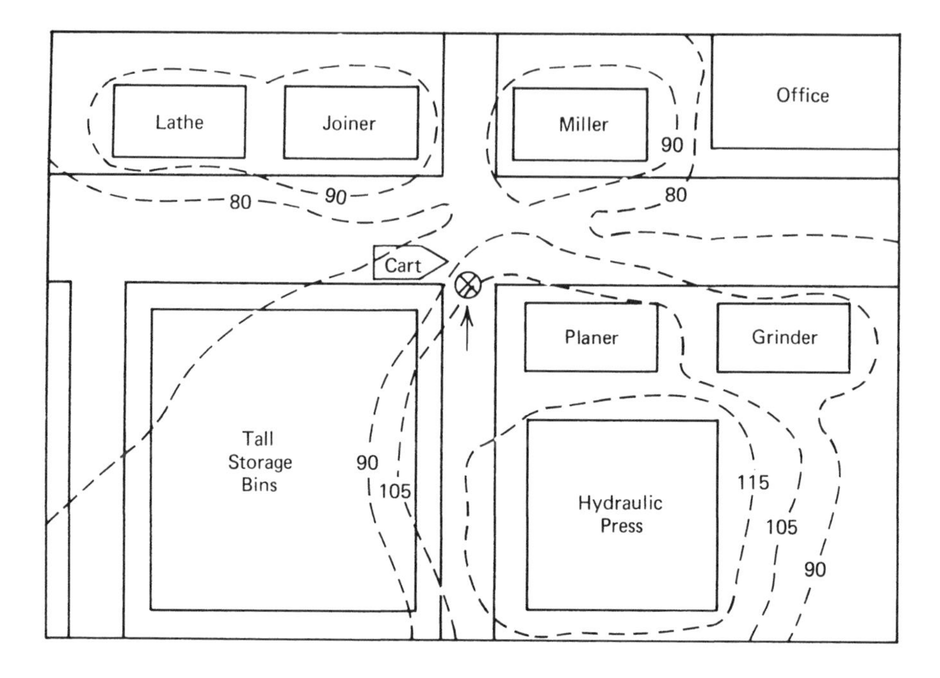

Figure 3.3 Gradient map of noise levels.

sharp increase in sound level in the mishap zone; noise here is loud enough to drown out warning shouts and the claxon sound. Because sound levels nearby are acceptable, only gradient mapping of sound levels would have revealed the problem. Similarly, problems with temperature and light might be made plain by gradient mapping of their intensities.

BACK TO NORMAL

The process of cleaning up and restoring things to normal after a mishap can in itself be dangerous. Take every possible precaution to keep anyone from getting hurt and to prevent more property damage.

The first rule is: Do not do anything until the mishap site is safe to approach. It is not safe if structures are weakened, chemical spillage is around, or other similar aftermaths of a mishap exist. If there is any doubt whatever, leave the cleanup and restoration job to experts. If your company is large enough to have an investigator, it is large enough to have skilled personnel to undertake a cleanup process.

If the cleanup task falls to you, expose as few people as possible to the activity. Anyone involved must be briefed on the possible hazards to be encountered in a new or difficult task. An old rule used by explosives people is appropriate: Keep the minimum number of people exposed to the minimum hazard for the minimum amount of time.

PUBLIC INFORMATION ACTIVITY

Nothing brings the news media faster than a mishap. Most of the time the investigator is faced with minor mishaps that do not involve the media, and it often comes as a surprise when public interest develops. When that interest does come, it appears when the investigator is facing difficult and complex tasks. The answer is to plan for it. Have someone picked, along with an alternate, to deal with the media. After that person is clued into the mishap action, let him handle things. Keep him fully informed so that he can handle the press, radio, and TV people.

You have seen skilled investigators bumbling their way through interviews with the media. The media cannot be ignored. They have deadlines to meet and news to get. The public deserves to be informed, and the media need to obtain information for them. If you do not provide news for them, they will use what they can get from whatever source can provide it. The image and reputation of your company are at stake. You are on the site and on the spot to give them that information.

Just as untimely is when your head office gets news of a mishap from some other source, perhaps a reporter trying to obtain background information, and hasn't the foggiest notion of what is going on. It is usually the responsibility of the investigator-in-charge-to keep the boss informed. Again, arrange for someone to carry out this task if you cannot do it yourself.

When the media are on the scene they will want pictures. You can seldom prevent them from taking them. To the contrary, you can make their job easier by pointing out dangers and trying to ensure their safety. A little help from you will improve your press image considerably. When possible, turn this task over to someone in the company trained in public relations, and concentrate on the investigation.

There is little reason for the investigator to deal with the media, irate citizens, public officials, or the like. Provide for it in the mishap planning document, and help should be at the mishap scene soon after you arrive. Until it arrives you must play the company diplomat and handle the situation yourself.

Let's assume that there is no one else to deal with the news media and, like it or not, you are on the spot. What to do?[11] The old rule is: Expect and prepare for the worst situation. Every crisis has its own set of problems, and your crisis is a mishap. Getting a number of calls from the news media may lead you to think the mishap is national news. Most likely it is purely local. An exaggerated response can be just as bad as underresponding. Know the scope of your crisis.

The investigator should not be faced with deciding who outside the company needs to know about the mishap. Company officials are the ones who should worry about who needs to know, how to reach them, and whether to hold a news conference.

When deciding what to tell the media, be honest and open, but don't speculate. Let them know an investigation is going on and they will be told when something definite is known. Whenever you know, they should be told the number and type of injuries, extent of damage, possible dangers to the community, and what is being done to prevent further damage and injury.

One voice should speak for the company. If necessary, an information center should be set up. If you are to deal with the news media, be accessible. Report your own bad news before they dig it out. Someone in the company should be taking the trouble to record the facts (including pictures and tapes) as they are uncovered so the company can tell its own side of the story. Update your story as it changes. Here are a few "don'ts." Do not:

- Say "No comment."
- Debate the subject.
- Attempt to place blame.
- Overreact.
- Deviate from company policy.
- Make off-the-record statements.

The best way to handle mishap public affairs is to have a coordinated plan worked out ahead of time and someone other than the investigator assigned to handle the news media.

SUMMARY

The first critical steps at a mishap scene can go smoothlv if good preplanning has taken place. At the site someone must take positive charge and proceed quickly, but with caution. Several actions should be taken almost simultaneously to ensure obtaining the short-lived and transient information. Locate witnesses, take statements, take first photos and make rough sketches, and return the scene to normal quickly. Throughout these early actions, assist the media as much as possible and keep management aware of what is going on.

QUESTIONS FOR REVIEW AND DISCUSSION

1 You have just had an industrial mishap in your plant. List your first three steps before going to the scene.

2 In the situation described in question **1**, what would be your first three actions as you approach the mishap scene?

3 To what extent should the investigator become involved in the fire-fighting and rescue effort at the mishap scene?

4 Suppose you are in charge of an investigation team in the mishap in question **1.** What would be your steps in getting organized through the first hour at the mishap site?

5 List five types of human witnesses who might not have been at a mishap site when the mishap occurred.

6 List five types of nonhuman witnesses that might be involved in a mishap.

7 What are the general guidelines for the first pictures to be taken at a mishap site?

8 Bring to class, *if you have one,* the type of camera that is recommended for general use in mishap investigation. Be prepared to demonstrate its ease of operation.

9 What is meant by the "standard four" in investigation photography?

10 Bring an example of a proof sheet to class.

11. Develop a 10-step outline to handle the news media in the event of a mishap at your workplace or school.

QUESTIONS FOR FURTHER STUDY

1 A truck from your construction company has bogged down in marshy land and tipped over, allowing a heavy lathe to fall off the truck. Name at least five recovery problems you face.

2 Give at least one solution for each of the problems associated with the situation in question **1**.

3 This morning one of your forklifts was found turned over in a far corner of your warehouse. The dead driver was pinned underneath. Where are you going to start locating witnesses?

4 List the first five actions you would take in the situation described in question **3**.

5 In the situation described in question **3,** explain how you are going to handle the news media. They are waiting for you in the front office with television equipment, and New York corporate headquarters is on the line. What are you going to say to each party?

6 You have hardly started your investigation, so you have little information to handle the situation described in question **3.** The news media claim you are withholding information from them. They intend to say that in the upcoming news broadcast. What do you do?

7 Locate an example of microphotography involved in a mishap and bring it to class.

8 Locate an example of macrophotography involved in a mishap and bring it to class.

REFERENCES

1 Ferry, Ted S., *Elements of Accident Investigation*, Charles C. Thomas, Springfield, IL 1978, pp. 24–25.

2 Bekerian, D. A., and J. M. Rolfe, "The Eyewitness as an Information Source," *New Direction in Safety*, Ted Ferry, Ed., American Society of Safety Engineers, Park Ridge, IL, 1985, pp. 216–229.

3 Sweeny, Joan, "Eyewitness: Seeing is Believing," *Los Angeles Times*, Wednesday, July 14, 1982.

4 Kuhlman, Raymond L., *Professional Accident Investigation*, Institute Press, Loganville, GA, 1977, pp. 79–123.

5 *Accident/Incident Investigation Manual*, Energy Research and Development Administration, Government Printing Office, Washington, DC, 1976, pp. 3-5 to 3-16.

6 Wood, Richard, letter dated Feb. 18, 1986, reference "Photography Section—Modern Accident Investigation."

7 Baker, J. Standard, *Traffic Accident Investigation Manual*, The Traffic Institute, Northwestern University, Evanston, IL, 1975, pp. 174–175.

8 *Fire and Arson Photography*, Eastman Kodak Company, 1977, ISBN 0-87985-187-2.

9 Donigan, Robert L., and Edward C. Fisher, *The Evidence Handbook*, Charles C. Thomas, Springfield, IL, 1965.

10 "Photographic Techniques for Accident Investigation" (35-mm presentation), Park Ridge, IL, American Society of Safety Engineers, 1987.

11 *When Every Second Counts—Crisis Communications Planning*, Western Union, Upper Saddle River, NJ, 1985.

PART TWO

Interacting Roles: Man, Environment, and Systems

The traditional roles of man, machine, and environment are discussed here in a nontraditional manner to facilitate an approach to mishap investigation that complements the rest of the book. The established contributions of the human being are covered, but environmental factors have been expanded far beyond the usual climate associations. While man himself is the supreme and most sophisticated of systems, Chapter 7 centers on operating hardware systems. This is properly reinforced by a chapter on the materials commonly found in operating systems.

CHAPTER FOUR

Human Aspects

It has been estimated that over 80 percent of accidental deaths result from the actions of individuals rather than externally imposed social or legal causes. Many mishaps occur when individuals circumvent protection that has been provided for them. Thus, a prime investigative concern is the focus on the human element. The investigator should always consider the involved individuals and their role in the mishap. Rather than arguing with those who tell us there is an element of human error in every mishap, we should instead redouble our efforts to find out if it is true. While we may not be able to change human nature, we can understand it better and resolve its contribution to mishap causation.

IMPORTANCE AND ROLES

The National Transportation Safety Board (NTSB) has investigated over 110,000 accidents in 25 years.[1] With few exceptions, the causes or probable causes were found. Considering the amount of human error identified in them, most of these mishaps could have been prevented by slightly different behavioral response or by simply anticipating the event.

Human errors are involved in nearly all mishaps. These may be errors of management, supervision, training, or of the workers themselves. Recognizing the errors is important if we are to reduce or eliminate mishaps.

The mishap investigator uses various tools, checklists, and methods to document relevant facts, conditions, and circumstances. For example, we have proven techniques and equipment to examine structures and engines. We are good at documenting hardware faults, and we enjoy a fair ability to record environmental factors such as weather, air samples, and terrain. However, we have not been nearly so successful in investigating and analyzing the human elements in mishaps. Unfortunately we will always be more successful in determining how and why the individual behaved in a particular manner than in anticipating inappropriate behavior that can lead to injury, death, and damaged or destroyed property.

One reason for our lack of success is that we apply the same expertise and methods used in hardware investigation to investigating human behavior.[2] We may even use persons trained in pathology and crash injury to investigate the "human factors" aspects. These techniques work for only a small number of the human elements. We do not have equipment to measure the intangible human factors and cannot readily draw conclusions from our investigation of these aspects. We should not expect investigators trained in pathology, crash injury, or hardware and/or environmental specialties to solve human aspects of mishaps any more than we can expect an average investigator to look into human factors.

Most of our investigative processes are based on deduction. We feel comfortable acting on the basis of fact. We believe that the validity of our findings is self-evident and is not open to challenge by our peers or supervisors. We can use scientific laws and measurements to evaluate our data. But when it comes to the inductive reasoning needed to deal with human beings we become uncomfortable, since the validity of our reasoning cannot be tested. Of course, our reasons are wide open to challenge and interpretation. We must deal with probabilities and likelihoods. We can be challenged and accused of second guessing or of drawing unfounded conclusions. Since investigation of human error largely involves inductive reasoning, our findings are often called speculative. We develop probable or acceptable conclusions instead of ones we can prove. Clearly, our techniques and thinking in dealing with human factors must be altered and greatly improved if we are to cite human causal factors with any certainty and credibility. There are no generally accepted techniques for the investigator to use in this area. Some definite approaches to verifying human factors in mishaps, however, are better than none. Guidelines for the investigator are presented in this chapter, but caution is advised for at least two reasons. First, no system is fully acceptable for getting at the human factors aspects of mishaps. Second, there is a widespread and valid feeling that only human factors experts should deal with the human factors aspects of mishap investigations.

UNSAFE ACTS

When an operator commits an unsafe act that leads to a mishap, there is an element of human or operator error. We are, however, decades past the place where we stopped there in our search for causes. Whenever an act is unsafe we must stop and ask Why was there an unsafe act? When this question is answered in depth it will lead us on a trail seldom of the operator's own conscious choosing. There are reasons for nearly every unsafe act and, given the resources, the investigator can usually find them. While their validity may be debatable, diligence will reward us with a reason for most human behavior. The "Human Hazards" section of Appendix A lists a number of factors affecting human behavior.

Ignorance, Memory, or Motivation

We so often hear the excuse "I don't know" or "Nobody told me" that it sounds like an errant child explaining his immature behavior. Granting that someone will sometimes do an injurious or damaging act on purpose, most people will try to do a task the way they think it should be done. We like to think that no one would deliberately cause a mishap. However, the mental model of a task may not concide with the actual situation encountered.

Assume that two or three things are possible when a person claims "I don't know." First is a lack of knowledge. He was never properly trained or instructed in a particular skill or technique to do the job he was undertaking. This leads to another type of human failure that involves supervision or management at some level.

Perhaps memory is at fault. It is possible that he simply did not remember the right way to perform. The sudden requirement to perform an infrequently assigned task does not leave time for practice and memory search so that it can be done properly. If you are suddenly placed in the driver's seat of a stick-shift auto after two years on automatic, you are in trouble. So is a machine operator faced with a similar situation. The reason for practice and recurrent training is to be prepared to meet task demands for seldom-performed tasks.

The human being is a marvelous machine when it comes to memory. While a little slow compared to a computer, human beings do excel at storing modest amounts of data for long periods and then recalling relevant facts at appropriate times. In computer parlance this is random access memory (RAM). It just takes a little while to make use of it. Humans have a long-term recall ability for generalized experiences but poor memory recall for sensory functions, particularly in the auditory mode. Their access time is slow compared to the computer, but they surpass in solving problems involving generalized patterns or prior experience. Once again it takes time. If immediate action is needed, there may not be enough time.[3]

If we fault ignorance or memory it is easy to say that the operators involved were not motivated to learn, to remember, to refresh themselves, or to really try to do a task the right way. A person neglecting to do something for his own reasons does seem to indicate a shortage of motivation. This reflects once again on supervision. Motivation is a particularly tough problem where operations are conducted continuously, perhaps rapidly and precisely, for long periods. The boredom that grows out of this type of operation can lead even the most well-intended worker into a mishap-producing situation. Machines can handle this type of operation much better than humans, and automation could well be included in the investigator's final recommendations.[4]

Few readers are in the business of investigating catastrophic aircraft accidents, but a summary of a 1979 accident[5] is appropriate here. The NTSB investigators found the cause to be the following:

> The Captain's decision to take off with snow and ice on the aircraft's wing and empannage [tail] surfaces which resulted in a loss of lift as the aircraft ascended out of ground effect. The pilot was not *motivated* to act as a professional and take the proper measures to ensure that the wings, stabilizing surface, and control surfaces are clean and free of ice, snow, or frost before he attempts a takeoff.

Another NTSB report in 1986 makes a similar point with another transport mishap. It involves a Panamanian-registered car ferry, the *L.A. Regina,* which ran aground on a reef near a small island with 215 passengers and crew aboard:

> The Board said the master [Captain] navigated "by the eye" instead of taking position fixes to check the vessel's progress. The master, who may have been drowsy because of medication he was taking, realized he was off-course when he saw the massive outline of the island seconds before the accident occurred. . . .
>
> . . . The probable cause of the grounding was the failure of the master to monitor the vessel's progress along the charted course by line plotting navigation fixes so as to detect the vessel's set and drift.

A truck driver could suffer the same lack of motivation to clean his icy windshield. A lathe operator might not clean up scraps around his machine, thus leading to a mishap.

Most of us can only process seven to twelve items in our short-term memory, and this limited capacity may keep us from handling a situation properly, despite the great motivation of avoiding injury or death.[6] Figure 4.1 is a model of the process using training to get the right response.

Attitude

Most unsafe acts can be attributed to the worker's attitude *and* to lack of knowledge or skill. Even after training to correct skill deficiencies, there is still the matter of attitude.

Some employees commit unsafe acts because of their attitude. Students of worker behavior can list several characteristics of poor attitude:

1. Poor work habits.
2. Indifference.
3. Daring.
4. Laziness.
5. Haste.
6. Temper.
7. Poor example.
8. Perfectionism.
9. Boredom.
10. Self-destructive action.
11. Peer-destructive action.
12. Superiority.
13. Inferiority.
14. Paranoia.

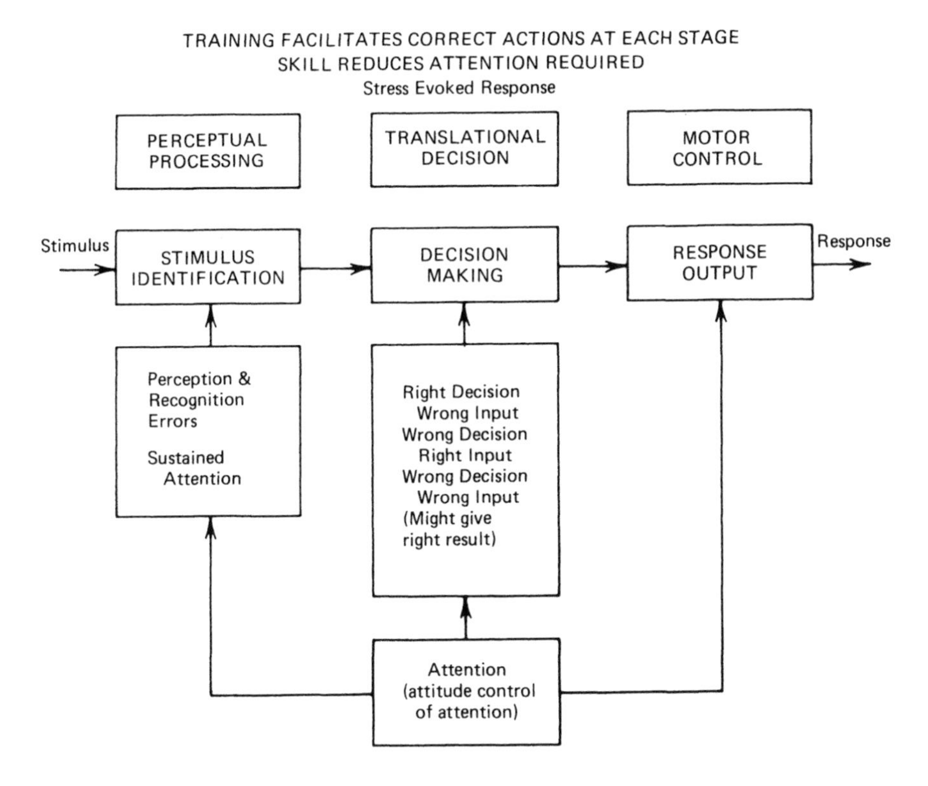

Figure 4.1 Training role in response acquisition (Hancock[7])

Inquiry into these characteristics will also reveal breakdowns in supervisor and/or management functions that need to be documented for more investigation.

It is common practice to identify an attitude as complacency when for no explainable reason a worker fails to properly perform a familiar task for which he is well qualified and has a mishap. Since he cannot offer a suitable explanation, complacency is often used to fill the void. The use of that term to define an operator's short circuit in behavior is open to question. Often the advice is "You're an adult now, solve your own hangups." The trouble with that is that the last person to recognize complacency is the operator himself. Complacency can be most readily recognized by others, particularly the direct supervisor. Failure to take corrective steps is an example of supervisory causal contribution. The investigator must realize that complacency tends to reduce the quality of both performance and judgment, an effect that is difficult to detect and measure.

Long experience has taught us that many mishaps happen because the average worker does not believe he will have a mishap. He thinks that it is always the "other guy" who gets hurt. If the worker doesn't have a mishap, his tolerance of the situation grows. The worker is pleased with the contentment that comes from being "fat, dumb, and happy": the signals of complacency.

One researcher[1] is firmly convinced that our cultural climate—that is, our aspirations and myths—is reflected in the prevailing attitudes toward machinery and that these attitudes contribute to a basic overall mishap rate. Supporters of this viewpoint believe the mishap rates would remain the same even if all specific measures of mishap prevention were realized. This opinion is not universal and has been contradicted by other research.

Errors

What do we mean when we speak of human error as a mishap causal factor? It depends largely on what classification of human error is being used. One process calls for breaking human error into errors of sensing, detecting, identifying, coding, and classifying. Another procedure divides errors into those involving sequencing, estimating, decision making, and problem solving.

When mishaps and unsafe acts are laid to human error, a more common classification considers errors of judgment, poor technique, disobedience of orders, carelessness, negligence, and errors involving signals below the threshold of the human sense organs. Contributing to these major classes are lack of experience, inadequate training, physical conditions, physical defects, and psychological conditions. These condition can be compounded by stresses such as the psychological—fear, grief, frustration, and anger—as well as the physical—stress, fatigue, temperature extremes, hunger.

It is obvious that the investigator faces a complex of factors that nearly defy organization. We know that operator performance varies a lot even during normal operations. A number of factors keep operator performance from being straight-line effective (Figure 4.2*a*). Operator demands also vary (Figure 4.2*b*) as new situations present themselves. When these two lines are superimposed (Figure 4.2*c*), a mishap is apparent when performance is degraded to a point where it is exceeded by demands on the operator.

A series of thorough investigations of aircraft mishaps over the last few years have implications that are easily transferred to any high-tech environment that depends on operator monitoring of controls, screen, dials, computer readouts, and so on. The NTSB has said about their findings and the role of monitors: "The increased automation has not necessarily reduced pilot work load . . . but has shifted it to monitoring tasks which the pilot has formerly had to perform, and there is evidence, from both research and accident statistics, that people make poor monitors."

Included in the research are several studies of human performance in relation to automatic systems. While its high visibility makes aviation a natural area for such research, we note several industrial mishaps in

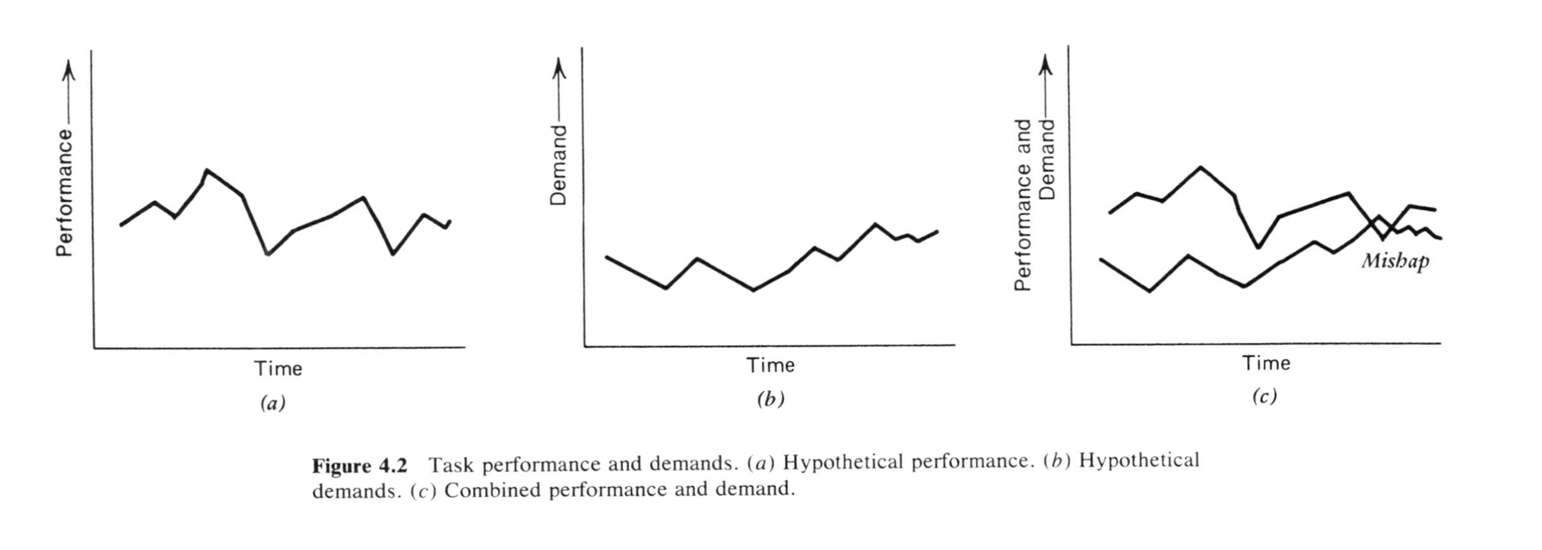

Figure 4.2 Task performance and demands. (*a*) Hypothetical performance. (*b*) Hypothetical demands. (*c*) Combined performance and demand.

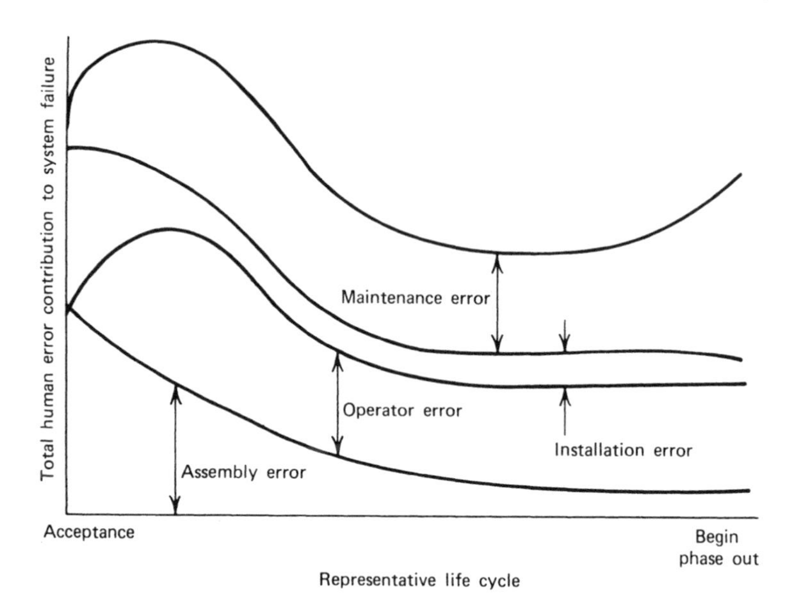

Figure 4.3 Errors in relation to life cycle.

chemical plants, nuclear facilities, and transportation that seem to have a logical link to the poor monitoring capability of humans.[8]

Conover[9] points out that in spite of these possibilities for mishap, humans are far better qualified to accept failure than a machine operating close to its optimal efficiency level. When the machine fails, the result is often catastrophic; that is, it no longer performs effectively if at all. On the other hand, humans are more effective over a sustained period, at perhaps 85 percent efficiency, for most tasks. Human failure is usually more gradual and the human system gives warning of impending failure through signs of fatigue and stress. If humans had to operate as efficiently as machines, they would also tend to fail catastrophically. To judge when our errors are most likely to occur, see Figure 4.3.

While human errors should not be excused as mishap causes, it helps to know they happen with a certain amount of predictability. Some basic error rates (BERs) have been compiled from a variety of sources. The BER is expressed as number of errors per million operations.

Task	BER
Read instructions resulting in procedural error	64,500
Installed O-rings improperly	66,700
Used wrong adjustment on mechanical linkage	16,700
Torqued fluid lines incorrectly	104
Installed nuts and bolts improperly	500
Machined (drilled and tapped) a valve to the wrong size	2,083
Omitted parts in a connector assembly	1,000

A NASA manual on collecting human error information[10] described a method for the study of human factors most involved in accidents. One section is a guide for interviewers who collect information on an event involving human error. To trace the actions of a person involved in a mishap it was hypothesized that certain behavioral events took place. Questions follow that can be applied by the investigator to better understand the reasons for the questionable behavior. The questions have been modified slightly for use here:

1 Was the necessary information acquired by the operator? Was the information correct? Was it in a format the operator could use in the time available?
2 Was the information properly evaluated?
   (*a*) With respect to quantity (enough)?
   (*b*) With respect to quality (consistent and reliable)?
3 Was the information properly processed, that is, did the operator reach an appreciation of the true state of affairs?
4 Did the operator select the safest and wisest decisions (based on the available information) from among the alternatives? If not, what other factors entered into his decision?
5 Was the decision effectively implemented once it was made?

To try out this process, consider the situation in Chapter 3 with the forklift and the hand truck. Place yourself in the position of asking the above questions of either operator. This will bring out many new considerations and focus on the human factors involved in the event.

Deliberate Actions

Some researchers believe that nearly all mishaps are unconsciously sought after. A few analytical psychiatrists view most mishaps as subconsciously purposeful acts, which, along with pain, damage, or loss, bring to the victim a variety of psychological gains and thus serve the psychic economy. One study of fatal mishaps came to the conclusion that if various events overload one's capability to cope, self-destruction becomes almost inevitable. This may seem remote, but it does not take much imagination to see how mishaps may satisfy some obscure need, including the need to be proved right, even when it hurts. Bruggink[11] gives an example of the driver of the family car who keeps telling his wife that he is going to be involved in an accident unless she quiets the kids down. "Some will find," he says, "more satisfaction in being able to say 'I told you so,' than in preventing the prediction from coming true." Examples of this situation can be found in many mishaps.

A study[11] of 300 industrial mishaps led researchers to the conclusion that most of the injured parties consciously or unconsciously sought the mishap and then consciously (deliberately) sought to preserve their symptoms. This cannot be validated statistically or proved legally, but this in-depth work should alert the investigator to consider the psychological possibilities.

Supervisory/Management Roles

Supervisors and managers may be reluctant to accept the thesis in Chapter 14 that nearly every mishap can be traced in some degree to supervisory and management roles. This is hard for some to resolve in regard to human error, since we have seen many of the problems in being able to define or control the human element in mishaps. A few examples of the management contribution are given in the hypothetical situation in the following paragraph. The supervisory/management implications is noted in parentheses.

An operator makes an error by selecting the incorrect action to carry out his operation and a mishap results. He was unable to match his actual action with those required by the task (training deficiency). The consequences of the action were not fully realized (supervisory deficiency). Appropriate training was not available due to cost and procedural problems (planning and budgeting). The operator made his error partly because the proper tool was not available (supervision, planning, budgeting, and staff support). Feedback on control of the action was poor and was not understood by the operator (staff, supervision, communication, and training). The operator made his error in carrying out a routine task that he had performed many times without a problem and that seemed too simple to warrant instruction (supervision). The task was carried out in a highly automated operation in which the operator was assigned to a boring, seemingly menial job that failed to challenge or stimulate him (staff and supervision).

The frequent references to training deficiencies above call for a bit of elaboration. Lack of proper training[12] is a factor often blamed on supervision and management. To grasp one aspect of this, visualize the life cycle of a machine as the horizontal axis of the graph shown in Figure 4.4. The vertical axis is the amount of operator training. The level of training is high when the machine is new, but as the life cycle progresses the training gradually declines. In reality, there are generally periodic small increases or humps as training is emphasized due to revised procedures or when an emphasis on training occurs. Generally, however, the line declines until the later part of the life cycle when the most equipment is in use and the training is almost nil. This occurs even with new operators and modifications continually being introduced to the same equipment.

It can be argued that as experience builds on the equipment there is less need for training. The training brings skill in the use of the equipment. Experience improves the accuracy of use once skill and ability have been acquired. However, the training need continues, since use of the machine changes, the machine is modified, and there are variables in the local operating environment. This leads to a situation that may require a level of

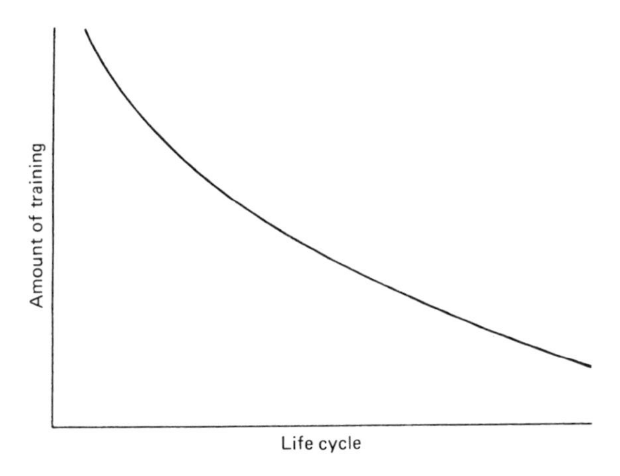

Figure 4.4 Training declines with time.

training as high as if not higher than the original level, at least in a recurring mode. Lack of attention to this area, even where the operators are well experienced, is a supervisory/management contribution to a mishap.

HUMAN LIMITATIONS

To understand how human limitations cause mishaps is also to know human capabilities and how they are degraded. We find that the human has a marvelous mixture of capabilities, but this combination has many ways to lower performance, sometimes to the extent that a mishap is virtually guaranteed. Human engineering deals with the industrial designer's role in helping humans succeed and enhancing their reliability. Failure in this role can lead to human failure. Stress, drugs, and alcohol make the human particularly susceptible to failure even when the designer's role has been fulfilled and the personnel selection process has been adequate.

Human Engineering

Human engineering, by one definition, applies psychology, physiology, medicine, anthropology, and engineering. The human engineering specialist represents the manufacturer, the machine builder, and the supplier. If he can also be pursuaded to consider the operator, he can make a large contribution to safety. The engineer, working independently, and alone, cannot know all about the human, even though he is professionally obligated to consider the characteristics, capabilities, and limitations of the human operator. In this regard the engineering world has not adequately accepted its moral obligation. At times, the machine has been designed to be so complex that only the most gifted could operate it, and then only with difficulty.

Table 4.1 Design Deficiencies Related to Human Error

Behavior	Design Considerations Identified
Lost hold on slippery article.	Surface friction not commensurate with task.
Underestimated or overestimated distance.	User was required to make critical estimate.
Placed foot under machine in familiar surroundings.	Necessary foot placement not provided for or made a matter for caution.
Used "wrong tool," which slipped.	Product not provided with proper tool. Wrong tool could be used.
Picked up hot metal from machine.	Not warned or trained. In wrong spot.
Gashed head when retrieving stock.	Machine had unnecessarily sharp corners.
Caught clothing on rotating handle.	Not warned about clothes and rotating handle as hazard under some operating conditions.
Misunderstood instructions on label.	Lable legible but instructions not clear.
Turned machine on instead of off.	Switch had to be pushed up for off position.
Spilled hot oil on self while cooking.	Appliance required both hands for simple operation. Should have allowed one-hand operation.
Burned on stove.	Off switch worked in reverse. Heat undetectable.
Child stuck fingers in metal fan.	Not properly guarded. Wrong material for home.
Fell on dimly lighted stairs.	Inadequate lighting and at wrong location.
Took wrong medicine.	Medicine label prepared for 20/40 vision.
Became complacent, struck hand in drill press.	Press guard not operational due to bolt failure. Press inoperative under these conditions.

As an example, consider the design of instruments. We realize that information displays should be legible, quickly comprehended, and require a minimum of attention and interpretation. However, engineering practice has for the most part aimed only at legibility. Neglect of the other factors leads to operator fatigue, reduced efficiency, and high error response. Catastrophes often occur as a result of just these factors and often involve highly skilled and experienced operators. Thousands of mishaps occur each year that do not make the headlines, all due in part to the same neglect.

Standardizing instruments is helpful but has not proved to be a good solution, because functional needs are always changing. Designs that fail to incorporate or leave room for adjustments to new functions will be inadequate and retard progress. Due to these functional changes we continually try to match the machine to the operator in terms of the new functions.

Our task as investigators is to identify where the engineer or designer has failed to take prudent actions to consider the possibility of human malfunction, thus setting the stage for human error. Table 4.1 gives some examples of design deficiencies as they pertain to human errors that resulted in mishaps.

Where safety is primarily governed by design solutions, the emphasis on behavior and motivation[1] is sometimes secondary. The situation approach to reducing human errors takes people as they are, not as one would like them to be, and recognizes that it is much easier and more permanent to change work situations than the people in them. In a well-run company there are enough procedures to assure compliance with established work practices. In addition, selection procedures can be controlled. Thus the situational approach appears most effective from a cost-effective standpoint. In practice the approach presents many problems.

A senate committee gave another viewpoint: "There can be no reliance on the margin for human error in air safety; safety must be assured through foolproof design." We could not apply this assurance in even our most liberally funded space programs, where attitudes and motivation are considered a key part of the program.

It would be counterproductive to safety for us to try to reduce operator failures by introducing complexity into maintenance or some other area. We would simply be trading off maintenance or some other kind of error for operator error.

Assuming there will be no sending error in the finest of warning systems and the best will in the world, an operator will occasionally do a stupid thing. Attitude ranges from casual acceptance of the environment to flagrant disregard for good procedures and safe operating practices. This seems to defy solution, but there are always ways to approach the subject. One way is through human engineering.

A report of the National Transportation Safety Board[13] cites a railroad mishap near Ramsey, Wyoming, where the probable cause was given as follows:

> The failure of the engineer . . . to comply with a series of restrictive wayside signals repeatedly by locomotive cab signals, including a "stop and proceed aspect" 6303 feet from the point of collision. Contributing to the mishap was the unauthorized muting of the cab signal warning whistle so that it could not alert the engineer when a more restrictive signal was passed.

A detailed understanding of this mishap is not needed to understand that an entire series of alerting signals were for some reason ignored and the train plowed into the rear of another. Furthermore, a cab warning device was shut off. This shows that the best intentions of designers may still be ignored and bypassed for reasons that will never be known or understood.

The ingenuity of designers is continually tested by human limitations and by the need for designers to judge what someone is going to do at some time in the future. The difficulty in trying to predict behavior may be illustrated by many cases. In one summary Peters[14] discussed the need for human research and cited several cases. A short summary of three will illustrate the challenge:

1. A woman locked in a lavatory attempted to get out by stepping on a toilet roll dispenser. It collapsed, and she was hurt. The owner paid for not foreseeing this use of the dispenser.
2. The landlord paid when he failed to repair water pipes and a tenant hurt her back bailing water from the bathtub.
3. When a housewife bumped a coffee pot against a faucet it shattered, injuring her. It was found that the manufacturer should have foreseen this possibility.

It is plain that the designer has a problem in preparing for many potential users who could be injured using the product. The investigator must often pinpoint the design deficiency, as remote as it seems, to meet the needs of the investigation's sponsor.

Let us consider some electrical mishap findings where it was found that lack of attention to basic principles contributed to the occurrence:

1. Failure to adequately enclose a device to isolate a hazard.
2. Failure to provide easily visible and accessible disconnection devices to cut off power from an enclosure.
3. Lack of warning lights to signal power was "on."
4. Failure to provide protective interlocks that "remember" when the human forgets.

For mishap-reporting purposes and for making corrective recommendations the investigator should recognize that the machine can excel in some areas. For instance, machines excel in monitoring both humans and other machines. The machine is best for performing routine, repetitive tasks, for doing precise operations, and in responding quickly to control signals. Machines can exert great force smoothly and with precision. Machines excel in performing complex and rapid computations with high-accuracy sensitivity and in doing many different things at one time. The machine is fairly insensitive to extraneous factors and can operate in environments hostile to humans or beyond human tolerance.

Just as we would not force a machine to operate close to its maximum limit, we should not force humans to do so. The investigator may readily assess the normal, safe, continuous operating range of a machine, but it takes all his skills and those of other specialists to determine when a human being has been overextended by deficient engineering design.

Human Reliability

Human reliability is so interactive with other aspects of human engineering that it is treated separately to make some points that may be of value to the investigator.[15]

Quantifying human reliability is an extremely difficult and perhaps impractical[9] task. Although there has been highly technical work in this field, and advances have been made, the application to everyday use is largely academic.

It is desirable, however, to identify relationships between human characteristics and certain factors that degrade performance reliability. The primary purpose of our discussion is to demonstrate how human error can be minimized through good design, thus increasing reliability. The investigator can then identify defects in the system under investigation.

Our human reliability ranges from 0.99995–0.99999 for simple, discrete signal responses to 0.970–0.990 for complex responses to complex inputs, including concurrence with another operator. This points out the relationship between probability of error and task complexity. Reliability ranges are considerably degraded if undesirable environmental effects such as the following are introduced into the task:

- Reducing the time allocated for task.
- Introducing abnormal temperature conditions.
- Subjecting the operator to shock, vibration, or severe oscillations.
- Failing to provide adequate or appropriate illumination.
- Imposing government restrictions on clothing and the like.
- Failure to provide compatible interface design.
- Introducing noise into the environment.
- Introducing stress through acceleration, vibration, humidity, muscle strain, and high noise levels.
- Imposing stress by confinement, isolation, or sleep deprivation.
- Introducing emotional stress through fear, anxiety, and boredom.

Human variability is such that performance reliability is subject to error from time to time for completely unexplained reasons. We should not increase this burden by inadvertently creating an unsafe error-producing design or environment.[16]

Techniques have been developed for carrying out a systematic analysis of the human role in interaction with the operating system. One has introduced the human factor into probabilistic risk analysis through a system of analytical trees, and in particular the Management Oversight and Risk Tree (MORT).[17] (See Chapter 11 for a full discussion of MORT.)

Stress

We are often told[18] of the harmful effects of modern living and the stress it brings to our lives, with resultant violence, often appearing as mishaps. According to behavioral scientists, continual anxieties arising from an accumulation of clearly minor incidents have in one generation produced increases in diastolic blood pressure, industrial mishaps, and the number of workers' compensation claims based wholly on job pressures. Unfortunately, those minor incidents can be more disquieting than major ones.

In the past 20 years there has been an average blood pressure increase of several points that has resulted in generally higher levels of tension. There is a certain tension level that often cannot be released legitimately, so it is discharged by displacement, sometimes as mishaps. There are several measures of this, but the one closest to us is the great percentage increase in the number of workers' compensation cases settled on the basis of cumulative job stress.

The study cited above[18] examines irritability or resentment in the workplace as a cause of industrial mishaps. The study reads as follows:

> Morale was low (in a plant), a high level of irritability existed and the guy who worked the machines felt like fighting. He didn't have anybody to fight with so he fought with the machines. Unfortunately the machines fought back. They fought back with accidents. People get so stressed up that they fight the best way they know and don't give a damn about the outcome.

Noise has already been widely studied in mishaps, but agreement is not complete on its roles. A study by the National Institute for Occupational Safety and Health (NIOSH)[19] showed that noise hinders communications and performance, annoys and frustrates, hinders sleep, and causes fatigue. According to this study, noise contributes to hypertension, psychosomatic disorders, neurosis, and acts of aggression. This finding, however, conflicts with some studies of the most stressful occupations, which again shows the need for caution. The NIOSH study showed that many people adapt to noise but show behavioral defects after the noise ends, such as impaired task performance and lower tolerance for postnoise frustrations. It also found that continuous noise to 110 dB does not have a significant effect on simple tasks but does have an effect where vigilance, sustained attention, and complex tasks are involved. High-frequency noise has a greater effect than low-frequency noise. Noise is a stress factor because of the annoyance it produces. The louder it is, the higher its pitch, or the more irregular it is, the more annoying the noise will be. Noise that interferes with communication or sleep is especially annoying.

We may be unable to describe the basic physiological characteristics involved in stress or measure it in biochemical terms, but we can gauge its effect on performance and judgment through human engineering studies. What is really important to the investigator is that noise affects performance as people become fatigued, their alertness declines, and the deteriorated stress condition produces mishaps.

Stress conditions are another of the unweighed factors that may not be well tolerated. Little research has been devoted to investigating human resistance to stress. While little is documented about the effects of combined stress, we can find signs of deteriorating performance in:

- Absenteeism.
- Mishap repeater behavior.
- Confusion.
- Difficulty in concentration.
- Spasmodic work patterns.
- Reluctance to change tasks.
- Generally lowered efficiency.
- Poor employee relations.

An investigator who finds evidence of those signs should look further into stress as a possible causal factor.

Medical Standards

Medical practice may suggest certain physical standards for workers, but they are often difficult to measure and their existence does not ensure compliance. As example, sound medical practice suggests that individuals with less than 20/70 corrected visual acuity in the better eye should not drive. There are hundreds of thousands whose corrected visual acuity will not meet this modest standard, yet they represent a large segment of automobile, industrial machine, and home product users.

In practice only a small percentage of the work force is examined in enough detail to reveal any but the worst handicaps. Most workers receive nothing that could truly be called a work physical before joining a company. The investigator should be quick to insist on a thorough medical examination for those having mishaps; although admittedly after the fact, it will aid materially in isolating human causal factors. A medical examiner especially trained for industrial practice should be used.

The situation is clouded by government regulations that preclude personnel offices from receiving some information they need to ensure a safe employee as far as physical condition is concerned. This comes about through "privacy act" provisions and the withholding of certain personnel records from anyone other than the holder and the person involved. The situation is further complicated by conflicting regulations and a trend toward keeping cumulative training records covering several places of employment.

Drugs and Alcohol

Almost any use of depressants, stimulants, or hallucinogenic drugs (see Appendix B) poses a hazard through an increased tendency to incur mishaps. Drugs taken for disease or bodily damage can present the same hazard. Some drugs are low in strength and pose no problem. However, many produce reactions that affect mental and physical performance to such an extent that a danger is created on the job. Some muscle relaxants and drugs for mental problems affect mental and physical stability. A larger problem is the lethal combination of various and often individually harmless drugs.

A special problem for business and industry is the lack of knowledge about drug use and the lack of control over those using drugs. Investigators often fail to consider the side effects on nonnarcotic, nonsedative, and nonstimulating drugs. Many produce lethargy, faintness, or sleepiness, and other side effects are often unknown. They are often used to treat a disease or medical problem, and the side effects are "the price we pay to treat the disease."

One study[20] failed to establish a direct relationship between the use of medication and accident rate. The small sample sizes used in this study rule out its industrywide use.

Alcohol is medically classified as a drug. Contrary to widespread belief, alcohol is a depressant, not a stimulant. It acts like an anesthetic and depresses the activity of the brain. It is addictive, and its contribution to mishaps is clear. Alcohol research has been extensive, and its contribution to traffic mishaps has been proved. Proving that alcohol is a factor in business and industrial mishaps is harder. In the controlled environment of the highway mishap, police are often authorized to perform breath alcohol tests on drivers suspected of drinking. Since alcohol is involved in well over half of the nation's driving fatalities we suspect a much higher involvement in industrial mishaps than we have been able to prove. The message to the investigator is to make this a suspect area and arrange for tests where alcohol is a possible factor. Note that where drugs are combined with alcohol the effects are often multiplicative, not merely additive.

Investigations show a high percentage of alcohol involvement in fire fatalities. One Maryland study showed alcohol to be involved in 206 of 516 fire fatalities. Some authorities estimate that 50 percent of adult fire fatalities are legally drunk. In Alaska this figure has been estimated at 80 percent.[21] Considering the wide use of alcohol and drugs of all types, the diligent investigator will consider the possibilities of both in any investigation.

Pathology

Although the coroner's office handles most details of a fatal mishap, the investigator has many things to do. He should (with the help of the coroner's findings) relate the mishap site and wreckage to the cause of death and injuries.

In a traffic or crushing mishap the investigator should determine the inhabitable area. It is possible that no one could have lived through the event, due to the drastic crushing or reduction of the area. However, if a body is found in an inhabitable space, he needs to know what caused the death. If no one could have survived in the area, interest in that death will be greatly reduced. If the operator was physically restrained in some way and was still badly injured or killed, then the restraint system is suspect and should be investigated. When we suspect the operator was killed by impact, we must be certain he was not previously incapacitated by disease, drugs, seizure, etc.

An autopsy is essential in mishap fatalities. The investigator is never authorized to do anything with a dead body. This, by strict law, is the domain of the coroner's office. A body can be moved only by prearrangement with the coroner. The autopsy is important to the investigator because it can tell him the cause of death. A man who falls into a machine and is mangled may have suffered a heart seizure, a blackout, or an electrical shock. Too often the investigator is looking for the wrong causal factor associated with the death. By the same token, if a causal factor is suspected, he can work with the coroner and voice his doubts and suspicions.

A postmortem examination should include alcohol and drug checks, but the investigator can ask the coroner to check other suspect areas. The result of the pathologist's examination of tissues and fluids may be the key factor in assessing the causal factors in a mishap. Proper taking and handling of tissues and specimens and their testing may mean the difference between knowing the actual causes of the mishap and having a "cause undetermined." Pathologists have been responsible for solving many mishaps. Since embalming rules out toxicologic results, the investigator's guidance to the coroner's office in suspect cases can be critical.

It has become common to refer to "psychological" autopsies with fatalities. This is a formalized approach to looking into anything with a psychological basis that might have led the human to a mishap. A former deputy director of the Bureau of Accident Investigation at the NTSB has spoken harshly[1] of looking into of the psychological events leading to a mishap, feeling that we are unable to justify these deductions for prevention purposes. Hancock[7] has also indicated the hopelessness of such findings for prevention purposes. Nevertheless, the psychological autopsy has moved ahead with large-scale testing in the military and has become a standard feature for certain mishaps along with the regular medical report.

SUMMARY

If there is an area of uncertainty in this book, it is probably this chapter on the human aspects of investigation. In spite of generations of research and libraries full of studies on the human role in mishaps, we still face largely uncertain areas. The dominant roles of the human being cannot be ignored, yet all too often we are investigating blindfolded, without a solid basis for our findings.

The reader should be aware of the many subjects and disciplines that bear on human failure and realize that there are complex and often unknown interactions among them. Simply being aware of the complexity of the situation offers some help for investigative action. In reviewing the headings for this chapter it is seen that each subject is, in itself, an area calling for subject matter experts. Even medical doctors and human factors specialists cannot possibly deal expertly in all medical or human factors areas. Being considered an expert investigator implies more that the investigator knows when, where, and how to get expert assistance than that the individual possesses expertise in all areas.

QUESTIONS FOR REVIEW AND DISCUSSION

1 The National Transportation Safety Board is mentioned at the start of this chapter, with the claim that in most of over 110,000 NTSB mishap investigations the causes or probable causes have been identified. Do you believe this? Why or why not?

2 Does the author make the point that an investigator trained in human engineering and physiological aspects qualifies as a human factors investigator?

3 Why are ignorance, memory, and motivation placed together in this chapter?

4 In the section on attitudes, mention is made of the advice: "You're an adult now, solve your own hangups." Give two examples of that situation where you work.

5 Look at Figure 4.2. Is this diagram appropriate to a lawnmower? To a kitchen stove?

QUESTIONS FOR FURTHER STUDY

1 Locate the investigation parameters within which the NTSB operates. Exactly what type of mishaps are they empowered to investigate?

2 Determine a BER for any operation in your organization where the BER has been generated in-house.

3 Look up the NTSB marine accident report cited in the section on unsafe acts, and bring in a more complete story of the event.

4 What is the relationship between reliability and quality control?

5 Give the exact medical standards required for your job. If there are none, state what they are for a friend's or relative's job.

REFERENCES

1 Bruggink, Gerald M., "The Last Line of Defense," *Legal Eagle News,* March 1975, Vol. 17, No. 7, pp. 1 and 6–10.

2 Schleede, Ronald L., "Application of a Decision-Making Model to the Investigation of Human Error in Aircraft Accidents," *Proceedings of the 1979 Seminar of the International Society of Air Safety Investigators, Montreal,* September 1979.

3 Klapp, S. T., and C. I. Irwin, "Relation Between Programming Time and Duration of Response Being Programmed," *Journal of Experimental Psychology: Human Perception and Response,* 1976, Vol. 2, pp. 591–598.

4 Schnieder, W., and D. Fisk, "Controlled Automatic and Visual Search," *Journal of Experimental Psychology: Learning, Memory and Cognition,* 1982, Vol. 8, pp. 261–268.

5 *Safety Information,* SB 79-65/2772, NTSB, Washington, DC, Aug. 16, 1979.

6 Miller, J., and R. Anbar, "Expectancy and Frequency Effects on Perceptual and Motor Systems in Choice Reaction Time," *Memory and Cognition,* 1981, Vol. 9, pp. 631–641.

7 Hancock, P. A., personal interviews, January–March 1987.

8 "People Make Poor Monitors," *Flight International,* July 26, 1986, p. 1.

9 Conover, Donald W., *System Safety Course—Man Induced Hazards,* University of Southern California and Man Factors, Inc., Los Angeles, April 1970.

10 Billings, Charles, et al., *A Method for the Study of Human Factors in Aircraft Operations,* NASA, Washington, DC, September 1975.

11 Hirschfeld, A. H., and R. C. Behan, "The Accident Process," *Journal of the American Medical Association,* Oct. 19, 1973, pp. 186 and 193–199.

12 Prendal, Bjarne, "An Attack on Human Factors as a Possible Cause," *Proceedings of the Annual Seminar of the Society of Air Safety Investigators,* Ottawa, Oct. 7–9, 1975.

13 *Safety Information,* SB 79-66/2632, NTSB, Washington, DC, Aug. 16, 1979.

14 Peters, George A., "Needed Human Research and New Methodology Considerations Being Forced by Law," Report 482, National Bureau of Standards, Washington, DC, July 1977.

15 Adams, J. A., "Issues in Human Reliability," *Human factors,* 1982, Vol. 24, pp. 1–10.

16 *Traffic Control and Highway Elements: Their Relationship to Highway Safety,* Automotive Safety Foundation and the Bureau of Public Roads, Washington, DC, 1963.

17 Nertney, R. J., and R. L. Horman, *The Impact of the Human on System Safety Analysis*, System Safety Development Center, U.S. Department of Energy, Idaho Falls, ID, 1985.

18 Dean, Paul, "Give Us This Day Our Daily Stress," *Los Angeles Times*, Oct. 14, 1979, Part X, pp. 1 and 20.

19 Smith, D. B., and Arnold Small, *Man, Environment and Systems Management*, Institute of Safety and Systems Management, Los Angeles, 1976, pp. 1–6 and 28.

20 Dunn, E. V., "Industrial Accidents and Medication: Is There a Relationship?" *Journal of Occupational Medicine*, 1979, Vol. 21, pp. 365–366.

21 Jones, Everett, "A Dangerous Duo," *Lifeline*, May/June 1979, pp. 28–29.

CHAPTER FIVE

The Environment

The environment is often considered to be everything around us. The term has its origins in the French word *environs,* meaning "around." Safety professionals have interpreted the word differently, depending on their backgrounds. Those with an industrial hygiene background tend to include subjects bearing on that area. Others like Grimaldi,[1] take a more specific viewpoint in considering health hazards. Wagner, in his widely used test,[2] uses the term in a much broader context, including nearly all the earth's surroundings: the earth itself, to include mining, agriculture, grazing, and so on; organics and inorganics in the environment; city environments; radiation; pollution; and social and population environments. The National Institute for Occupational Safety and Health has chosen to concentrate on the industrial environment[3] and also considers many areas peripheral to the environment.

The safety professional is interested in nearly all types of environments. He is, for example, increasingly called on to investigate mishaps that involve hazardous materials and hazardous wastes. Without meaning to slight the many fields of environmental concern, we have chosen to dwell here on five areas[4] that include the social, technical, economic, and political/legal environments, with added emphasis on the problems generated by management structures. They include communications, weather, noise, and terrain. Appendix A lists hazards considered in this approach.

The need to examine our environment for mishap investigation was precisely stated by Lederer[5] when he said, "We have been dealing with the poisonous fruits and not the seeds of 'mishaps' or 'mishap causations' stemming from human, social, political, or economic frailties of our way of life."

SOCIAL ENVIRONMENT

As mishap investigators, we are deeply concerned with what people learn and accept, which are facets of the social environment. Our concern for human behavior in mishap causation points to a need to understand how languages, customs, attitudes, habits, and values interact with the social environment. Consider a few examples:

1 In a U.S. factory the work force is largely Hispanic. As an overhead crane swings dangerously close, a warning is yelled in Spanish. An English-speaking worker does not get the word and is badly hurt.
2 In a few countries, the "stop" traffic signal is ineffective—it is the local custom to ignore such signs. A U.S. truck driver suddenly stops for a signal and is rear-ended at a good speed by a local taxi. Since the truck driver is driving your company truck, you must investigate. In the eyes of the local authorities your man is considered obviously at fault.
3 A young student is a part-time worker for your plant during the summer. He does not seem to grasp the need for maturity and care in driving your delivery truck. While "burning rubber" out of your plant entrance, he pulls in front of a fast-moving truck and "totals" your truck. Not only is he casual about the whole affair, he does not understand your concern since you are covered by insurance.

In the last example the values of the young driver will probably change in time as do all our values, but that might not translate into desired, mature, safe behavior. On the contrary, a good dependable worker may, for instance, gradually become disenchanted with his job and bosses. The result may be disinterest, antagonism, or sullenness, all symptoms of mishap causes. Last week's ideal employee can become this week's leading candidate for a mishap investigation, all because of changing values and attitudes, a regular part of the social environment.

Social values are constantly changing. A few years ago some minorities would not have been accepted in some workplaces. They are now accepted and often bring new problems with them. In the past, some workers would have been so upset abount minority coworkers that mishaps would result. Now the situation may be accepted, but it may create new hazardous situations. Note that the solution to one problem of changing values does not necessarily work for other similar situations, even in the same place. A grudging acceptance of blacks in one department cannot assure acceptance in other departments. Consider the entrance of women into the plant work force. At this writing,[6] the Supreme Court has just ruled that not only will women be guaranteed an appropriate leave, they will also be given their same jobs back. The possibilities for mishaps arising from a person returning to a particular position after an absence of several months are dramatic.

There are changing social pressures from groups in and out of government who insist that certain hiring quotas or standards be met. Workers who enter the workplace without special training may receive dangerous work assignments not expected or tolerated by older employees.

Pressures arise from the suspicion that workplace hazards exist, even when the danger is not known for certain. Sometimes research suggests the possibility of cancer or other disease at a future date, and although the data are meager, they are enough to disrupt or demoralize the work force with resultant mishaps. On the other hand, proof of danger has been shown in many workplaces as radiation and exposure to inorganic metals such as mercury, lead, cadmium, asbestos, and the like. Organized social pressures may attempt ot correct such working conditions but will often result in a new set of work dangers that can lead to mishaps. For example, the use of bulky protective clothing can result in clumsiness around moving machinery. Then there are social pressures from labor and peer groups. Failure to align with fellow workers in an organizational attempt may bring ridicule and peer pressure, causing nervousness and apprehension on the job, which are reflected in more mishaps.

Some of our lifestyles reflect an inordinate concentration on "doing one's own thing" to the exclusion of care and concern for fellow workers. This is often reflected in higher mishap rates. A challenge exists for the mishap investigator who is trying to uncover and reconcile the social agencies that interacted to cause a mishap and who must then recommend protective action to prevent recurrence.

Another social problem exists in the varied age groups represented in mishaps. The Surgeon General[7] has reported that mishaps rank as the most frequent cause of death up to the age of 40. The highest death rates for mishaps occur among the elderly, where the death rate is twice that of young adults. Fifty percent of the deaths among children 1 to 14 years of age are due to mishaps; a 50 percent reduction in this rate would meet national goals[8] on mishap deaths per 100,000 population. Because the social values of these three groups—youngsters, young adults, and older adults—differ, that our investigations should probably be different for the different age groups involved in mishaps.

Lead poisoning is a striking example of an environmental hazard for the young, particularly those living in older neighborhoods where lead-based paints often remain on the walls. Elevated levels of lead in the blood of as many as 24 percent of those children suggest the need for specific corrective actions far different from an investigation of older adults for the same diseases.

In another age-related consideration, the older driver has fewer severe mishaps but is more likely to be killed. Also, because of deteriorated performance, older drivers have different types of mishaps. Over 65 years of age, the chances of being killed as a pedestrian are three times as great as under 65.

We should consider that 4 percent of the males between ages 66 and 70 are employed. This represents 6.5 million people, and the number increases yearly, due in part to recent legislation. On average, older people are smaller, their vision is weaker, they have greater sensitivity to glare, a lower degree of dark adaption, and a narrower field of vision. The vision decrease is not trivial, but the lost peripheral vision is more of a mishap factor. As people age they often develop problems with attention span and are more easily distracted. This presents problems in maintaining skill levels. Falls, involved in over 50 percent of the mishaps among older people at work and play, have a special role in their environment. These factors bring new investigation challenges and new opportunities in mishap prevention. Designing safe work and work environments for the oder worker calls for discarding many of the accepted workplace standards.

TECHNOLOGICAL ENVIRONMENTS

The ever-increasing pace of technological developments (see Appendix A) intrudes directly into the workplace, where each change brings new mishap situations and conditions to be investigated, often for the first time. This is particularly true of the new sciences and places where exotic materials are processed and new chemicals are used.

The technology explosion has also brought new investigative tools. The computer, for example, can be used to conduct nationwide searches for similar and parallel events. It can be used to diagram or analyze mishaps, or to enter a data bank for information on similar situations and their solutions. It can compute the probability of an event taking place. Computer graphics have become so common and the software so reasonably priced that mishap situations can readily be brought to the video terminal. New tools allow us to see things we did not even know existed. We can now model a mishap without even being close to the wreckage and see the results under controlled simulation.

Unfortunately we cannot turn to the tried and proven techniques of the past to apply our new solutions. The technology explosion has outdistanced our investigative ability. This will increasingly require the use of specialists in investigation. We must accept this and realize that modern technology does not allow one or even a few persons to know everything. Technology has allowed us to do more things more quickly and often more safely, but it gives us a whole new set of hazards to be protected against and investigated.

ECONOMIC ENVIRONMENT

Each year 100,000 American workers die from occupational illnesses, and another 400,000 new cases appear. Work-related injuries kill about 11,500 workers each year and affect 9 million more so severely that they require medical care or are at least temporarily unable to work. The cost of these mishaps has already been discussed. The advancements of inflation and the advent of new investigative tools have made good investigation possible and necessary but perhaps a luxury. The investigator who must explain to a cost-conscious management the cost of a thorough investigation as well as the need for such investigation requires more than mere investigatory skills.

Tight money reduces investigative funds at the same time there is more need for good preventive efforts, which should be based at least in part on a thorough investigation. The message is: We need better investigations using the latest techniques. We must get the most usable information at the least cost without jeopardizing the investigation results. Justification for good investigation must be worked out as higher payoff, fewer mishaps, and increased operating efficiency.

This situation is compounded by an excellent mishap experience in some companies; good mishap rates are more difficult to reduce. An increased effort just to stay even or to attain modest reductions will not appear as cost-effective to management unless they are extremely farsighted. Few will say it, but the tendency is to accept some level of mishap loss as a cost of doing business. Management may be sympathetic to an intensified investigative effort, but the effort may not seem worthwhile.

Management can visualize something like Figure 5.1 and may see position *P* as the point best suited for them. Or, in spite of the higher total cost of mishaps, they may see the ideal point as well to the left of point *P* on the total cost line. In that case the results are twofold:

1 Management is willing to risk more mishaps so they can pay less for prevention and place the resources elsewhere.
2 The funds available for investigation and prevention will decrease.

It falls to the safety staff to convince management that a better investigative effort is good business.

The economic environment can redirect our efforts in many unexpected ways. For example, a devalued dollar and an energy crunch may result in new fuel and transport modes. The investigator may be called upon to investigate motorscooter and bicycle mishaps instead of car or truck mishaps. Carpooling leads to more people being involved per mishap. Alcohol or alternative fuels may replace gasoline. Smaller and lighter cars mean more serious mishaps.

One solution to this myriad of problems is better investigation with the same resources. This is accomplished by knowing precisely what to look for, how to go about your job in an efficient manner, and what techniques to use for the resources invested. Some of the techniques covered in Part III of this book are useful in that respect even for the modest, occasional investigation.

POLITICAL/LEGAL ENVIRONMENT

The legal aspects of investigation are so important that a separate chapter (Chapter 18) is devoted to the subject; a book would be more appropriate.

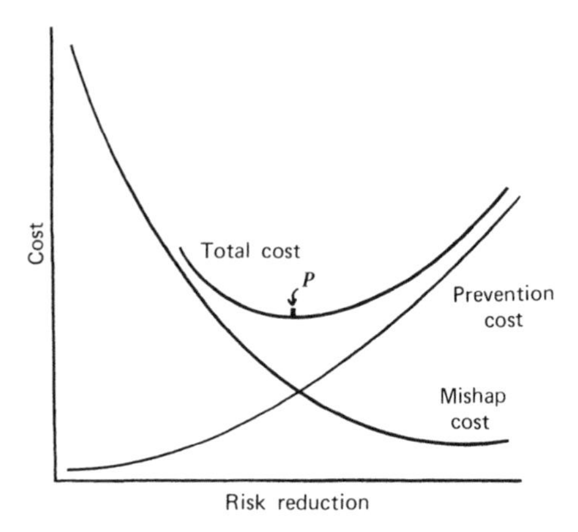

Figure 5.1 The cost of mishap prevention.

By political/legal environment we mean the effects of legislation, often that inspired by social and economic interests. For example, the Equal Rights legislation brings women into the workplace in unaccustomed positions and levels. We now find women in positions not fully analyzed for them, doing tasks designed with a strong man in mind, and sitting in operating positions scientifically prepared for the average male. Another example of legislative changes is the removal of 65 years of age as the mandatory retirement age, placing more older workers in regular workplace positions.

Our work force enjoys more protection from harm because of landmark court decisions. The employer must now face safety issues never seriously considered before. Even the average consumer is forcing investigative processes not seen before. Products-related mishaps cause about 30,000 deaths and 37 million injuries yearly. As a result, the Consumer Product Safety Commission brings pressure on the manufacturer to market a safe product. Investigators are now concerned not only with conserving company resources but also with providing protection against consumer-generated lawsuits. Safety managers find themselves in investigative roles for the mishaps suffered by company suppliers so that supplied products will meet the requirements of their own final products.

The definition of an accident or mishap is still in limbo with respect to cumulative trauma, be it stress, carcinogens, or whatever. With the aid of specialists, the investigator looks at such events, or the possibilities of them, sometimes decades after the onset of the event. The true dimensions of the asbestos hazard, for example, were only recognized after more than 30 years. People in many other occupations, such as rubber and plastic workers and those in some coke oven jobs, are showing significantly higher cancer rates than the general population.[3] There is obviously much more to come in the cumulative trauma area.

A most important piece of legislation has been the Occupational Safety and Health Act (OSHA), which not only forces investigation (not necessarily good ones), but imposes severe penalties where neglect is present. These penalties increasingly take the form of criminal proceedings with fines and jail sentences. This forces the investigator to do his job with a thoroughness that assures proper preventive action and penalty avoidance.

The new look in safety also involves management and staff to an extent not considered before in terms of personnel selection, purchasing, management, and supervision. It has even brought penalties to the safety manager for inadequate investigation and corrective actions.

Recent legislation bars access to certain employee information. That allows people into the work force who in the past would have been screened from consideration for employment. Good investigation will help us to deal more effectively with this situation.

MANAGEMENT ENVIRONMENT

There was a time when the main investigative focus was on the machine, looking for ways it had failed or in which an operator had contributed to the failure. Current approaches focus on management. A trend to look for the ways management pressures contribute to mishaps is emphasized in a separate chapter (Chapter 14) and in parts of other chapters that address the management environment.

Careful examination of any mishap will reveal several management factors that made the mishap possible or could have prevented it. Sometimes the mishap results from an omission of management. The pressures on management are often external to the operation. Legislation for product and worker safety, the social and political pressures mentioned earlier, and the influence of special pressure groups have brought many mishap-contributing factors into play. The internal influences of labor, profits, stockholders, and top management contribute to subtle pressures causing mishaps, often delaying investigations, and sometimes pressing for more thorough investigation. These pressures, often subtle, do not always support investigation and often counter it. The need to compete in the marketplace often forces management into using new production processes they have not fully examined in terms of preventive action and that are more expensive to alter after mishap investigations reveal deficiencies.

The contributions of supervisory management can be seen through examination of a mishap. Take the case discussed in Chapter 3, of the forklift running into a handcart on the loading dock:

1 Supervision is evidently a factor, because the job supervisor was not even monitoring the activity.

2 Supervision is seen to be a contributing factor because defective equipment was allowed to operate (no light on the forklift) and a rider was allowed on the forklift.
3 Supervision is seen as a factor in that noise was allowed that caused verbal instructions and warnings to be drowned out.
4 Management contributions are recognized in that:
   (*a*) The loading dock was not properly lighted.
   (*b*) Defective equipment was permitted to operate.
   (*c*) Ineffective supervision was the rule of the day.
   (*d*) The mirror did not permit night viewing.
   (*e*) A downward sloping ramp did not permit adequate control of heavy handcart loads.
   (*f*) An obviously sloppy mode of operation existed.

These factors could all have been corrected with little cost and effort. They point to a generally poor management environment. The results of investigating this one mishap would, without question, reveal many more management oversights that signal a generally poor and costly operation.

Although this section indicates that management emphasis is a peaking trend, it is only part of a much more encouraging trend, that of looking at *all* factors contributing to a mishap, not just the hardware, operators, regulations, and so on, but all factors that were involved. (Evidence of this growing trend is seen in Chapter 12.)

COMMUNICATION PROBLEMS

Communication problems as discussed here in an environmental context refer to barriers to communication that permit mishaps to occur. Written, visual, and oral communications are at the heart of any type of operation. Open communication generally results in a better working environment. Thus, a poor mishap record usually reveals communication deficiencies.

Failure by a worker to understand or know something about his work environment is a communication failure. If the company safety policy is not being carried out, that is a communication failure. Such failures need not involve an equipment operator in a mishap. There may have been different failures at various and higher levels that have made the mishap inevitable. Perhaps a worker or supervisor was not told of a new or revised procedure. Maybe management did not keep itself informed and up-to-date and thus permitted a mishap environment to develop.

It is not unusual for a mishap to occur as a result of the action of a person not involved in the operation at all. Consider an electrician working on a motor. He turns off the electrical switch and properly placards it. He then leaves the area to secure some tools. During his absence a worker needs power. After looking for signs of activity he concludes the job is complete and that the electrician forgot to remove the placard. He removes the placard and turns on the power. When the electrician returns he is badly injured by moving machinery he thought was deactivated.

Let us return to our familiar loading dock mishap for a look at communication problems:

1. The supervisor was away from the immediate scene and was therefore out of communication with the parties involved in the mishap.
2. Due to the noise of a truck engine, warning calls and yells could not be communicated.
3. Due to lack of lights on the dock and forklift, certain warning information could not be transmitted.
4. Information was not given to management about certain operating deficiencies at the site, or if it was, the failure of a correct response (feedback) shows a lack of communication.

A few basics about safety communications are worth noting as probable cause factors:

1. Most people believe that a communication requires feedback; that is, when a message is sent there must be some sign that it was received and understood.
2. Communications tend to be distorted in proportion to the number of channels they must pass through. A verbal order from top management, relayed through several channels, will surely be distorted by the time it reaches the operator.
3. The greater the number of communication channels being used at any one time, such as in a meeting, the greater the probability of a communications breakdown.
4. Communications are not all good, even when they are well-intentioned. Telling someone to "watch out" for something may be interpreted as telling him how to do his job, in which case it will be resented. The communication may be so distorted in passing through channels that it results in the undesired event being carried out. A worker may twist the meaning of what is said because he distrusts or dislikes the person trying to communicate. Defensive behavior of this kind endangers effective listening. It prevents the listener from concentrating on the message, and what he receives is distorted. In supportive behavior the opposite is true.
5. In a very noisy environment, say 90 dB, a person is only half understood when using a loud voice only a foot away. Many industrial areas such as a turbine hall or steam valve area exceed this level. Around a punch press we could expect over 100 dB. If the noise varies

in intensity it is far more of an irritant than a steady loud noise. After one study[9] it was concluded that:

(*a*) Increased noise results in a longer response time.

(*b*) The number of errors grow as noise level increases.

(*c*) The number of errors per production unit increases with noise level.

(*d*) Intermittent noise reduces the individual's capacity for quick and precise execution of coordinated movements.

On the subject of detectability of audible warnings, an earlier study made the following observations:

(*a*) The uniqueness of an audible warning system is of no particular significance.

(*b*) To elicit a response the detectability level must be much higher than for simple audibility (an estimated 10 dB on an order of magnitude).

(*c*) The type of background noise is important in determining detectability level. A loud horn may not be heard at a busy corner where there are other loud horns and much traffic noise.

(*d*) Intersection collisions with emergency vehicles may be laid to a simple failure of a converging vehicle to detect the emergency vehicle siren.

6 The communication may not be well presented. There is a need for clarity, brevity, and accuracy in safety communications. The skill to do this must be learned. In our mishap environment, verbal communication skills may be at fault, or the written word may be misinterpreted. Noise or hearing loss can cause a misunderstanding. Perhaps no reply is received to a request and is taken as an affirmative answer when the question was not really well understood. Indeed, the lack of an answer may simply indicate that the approving authority wanted time to think about the request.

Communication is the main tool for effective behavioral change in work relationships. Since it is so necessary for safe operations, and since so many of us are poor communicators, we must consider the communication environment in mishap investigations.

WEATHER FACTORS

The implications of weather as an environmental causal factor in mishaps seem straightforward. If we examine the weather elements—wind, rain, lightning, fog, snow, ice, floods, and so on—we can visualize how they made this mishap possible, or impossible to avoid. Think of the weather in terms of improper performance or restricted visibility. A few examples place the weather's role in perspective:

1. Adverse weather conditions may be present and still not contribute to a mishap. Suppose a machine operator, indoors in a plant, bypasses a machine guard and gets his hand caught at a nip point. The fact that it is pouring rain outside and the streets are slippery does not affect the mishap and normally would not be a cause factor. If, improbably, his machine and operation were outside and exactly the same thing happened, weather might still not be a causal factor, even in the downpour. If he could not see because of the rain, *then* it might have been a causal factor.
2. Consider the drunken driver on an icy street who runs a red light and steers into a utility pole. If that is all there is to it, weather might not be a factor. However, if he tried to brake and slid on the ice into the pole, then weather was a factor.
3. There is a sound basis for not considering weather as a mishap causal factor if the weather was accurately known and the hazards associated with it were recognized. For example, if a taxi dispatcher warned a driver of icy conditions on a certain stretch of highway and the driver then proceeded on the road at an unsafe speed to a mishap, the icy condition is a questionable cause. The driver chose to ignore the known hazard, and emphasis should be placed on the behavior instead of the weather.
4. In a final example, consider the pilot flying a twin-engined plane in clouds. An engine fails because of a bearing problem. If the pilot shuts down the bad engine and continues the flight to a successful landing, then weather is not a factor. If, in landing, the plane cracks up due to cross winds and surface ice, then weather becomes a factor. If it crashes because the pilot could not maintain flight speed on one engine, weather might not be a factor at all. If its wings iced up and the plane was not able to maintain flying speed, then weather does become a factor.

A major concern about weather as a factor in mishaps is how to determine the extent to which weather affected the event, that is, how much more damage or injury was due to weather? Did the weather present delays in rescuing people and getting them to medical care? Some, with justification, say that recovering people and machines may be the most dangerous part of a mishap. The investigator is often charged with planning for these eventualities, so his work is often directly reflected in the rescue and recovery effort. You may find yourself as the investigator-in-charge, and, as such, responsible for the recovery of damaged equipment. In adverse weather conditions, your task is often made many times more dangerous. Three things should be remembered to minimize the dangers present:

1 Whenever possible, leave the recovery operation to those who specialize in such events.
2 Keep everyone clear of the recovery operation who does not need to be involved.
3 Take special care if adverse weather conditions exist.

TERRAIN CONSIDERATIONS

Terrain is a concern in many types of outdoor mishaps. Most terrain elements can be noted in a quick drive on a dirt road through hilly country. Consider the visibility restrictions of hills and rocks, the rough surface of the roadway, the steepness of the climb or descent, the narrowness of the road, and other terrain factors. The vehicle in our mythical mishap may have gone over the side of a steep hill; the subsequent rough fall caused most of the damage; the steepness of the hill makes rescue difficult and perhaps makes the wreckage undetectable from the road.

Terrain may be a factor in such a simple action as a person walking across a storage yard, a pebble causing him to turn his ankle and fall. This situation will not likely warrant an investigation unless he was helping to move delicate equipment or was carrying a jar of nitroglycerin.

SUMMARY

The role of environmental factors in mishap causation is admittedly given scant review here if environment is classically defined as "all that is about us." In fact, several college courses are devoted entirely to the subject of environmental safety, and degrees are awarded in the subject area. This chapter covers only a few of the environmental factors that can be involved with mishap investigation. Others are covered elsewhere in the book. The skill of the investigator is often taxed to determine if the environment should actually be noted as a causal factor in a mishap.

QUESTIONS FOR REVIEW AND DISCUSSION

1 Discuss the effectiveness of the term "environment" as defined at the start of this chapter.
2 Which of the major divisions in this chapter do you consider most important to the investigator? Why?
3 Give an example of a mishap in which the social environment was clearly a causal factor.

4 Give three examples of things you might wish to consider in investigating a machine shop fatality involving a 69-year-old employee that might not be considered if he were half that age.

5 In the organization you or your father or mother work for, describe the management attitudes toward thorough mishap investigation. Cite solid reasons (interviews, practices, memos, etc.) for this attitude, bringing in related materials if they are available.

6 Cite three government agencies or pieces of legislation not mentioned in the text that are an environmental influence on mishap investigation.

QUESTIONS FOR FURTHER STUDY

1 Look up three definitions of environment outside of this text, and discuss their application to investigation in one paragraph for each.

2 Give three examples of the technology explosion that have aided the progress of mishap investigation techniques. Cite your references.

3 Take any NTSB accident report (not a publicity release) and cite all the technological and economic environmental factors involved. Bring the report to class with you.

4 Take any NTSB accident report (not a publicity release) and cite the weather and terrain factors involved. Bring the report to class with you.

REFERENCES

1 Grimaldi, John V., and Rollin Simonds, *Safety Management*, Irwin, Homewood, IL, 1984.

2 Wagner, Richard H., *Environment and Man*, 3rd ed., W. W. Norton, New York, 1975.

3 *The Industrial Environment: Its Evaluation and Control*, Governmant Printing Office, Washington, DC, 1973.

4 Ferry, Ted S., *Elements of Accident Investigation*, Charles C. Thomas, Springfield, IL, 1978.

5 Lederer, J., "Methodologies and Patterns of Research in Aircraft Accidents," *Annals of the New York Academy of Science*, May 1963, Vol. 107.

6 *Los Angeles Times*, Jan. 14th, 1987.

7 Richmond, Julius B., "Surgeon General's Report: Prevention is the Key to Solving Occupational Risk Puzzle," *National Safety News*, October 1979, pp. 43–47.

8 Smith, D. B., and Arnold N. Small, "Elderly as Cause and Victim of Accidents," *Proceedings of the 1975 Human Factors Annual Technical Meeting*, Dallas, 1975.

9 Grimaldi, John V., "Noise and Safe Work Performance," from the study "Sensory Motor Performance Under Varying Noise Conditions," *Ergonomics*, November 1958, Vol. 2, No. 1.

CHAPTER SIX

Materials

INTRODUCTION

Material failures are in many respects like most other failures in their basic causes. It is difficult to conceive of any failure or difficulty, including that of materials, that does not fall into one of these classifications[1]:

1 Ignorance.
- (*a*) Incompetent people in charge of design, construction, or inspection.
- (*b*) Supervision and maintenance by personnel not having the necessary knowledge.
- (*c*) Assumption of vital responsibility by personnel not having the necessary knowledge.
- (*d*) Competition without supervision.
- (*e*) Lack of precedent.
- (*f*) Lack of sufficient preliminary information.

2 Economy.
- (*a*) In first cost.
- (*b*) In maintenance.

3 Lapses or carelessness.
- (*a*) An otherwise competent person shows negligence in some part of the work.
- (*b*) A contractor or superintendent takes a chance, knowing he is doing it.
- (*c*) Lack of proper coordination in production of plans.

4 Unusual occurrences—earthquakes, extreme storms, fires, and the like.

However, we are going to look at failure differently in this chapter. We will look at how materials actually come apart, collapse, disintegrate, dissolve, and so on, even though their basic failures may be found in the above list.

Look around the space you are now in and note the different materials that make up the structure and furnishings. In the ordinary room you will see cloth, plaster, glass, wood, plastic, steel, perhaps brick, cement, brass, aluminum, and perhaps a host of other substances. Energy has been required to produce these materials and set them in place. It is this locked-up energy that concerns you as a mishap investigator, for as Haddon[2] points out, a mishap is a "spill of energy across some boundary, personal or societal." When a material fails, it generally releases some kind of stored energy by losing support, fragmenting, causing additional failure, or releasing some toxic substance through combustion and/or chemical reaction. Determination of the cause of the initial failure requires some knowledge of the physical and chemical nature of industrial materials.

Humans have shown a steady inclination to remake their world by adapting the materials around them in increasingly complex ways for protection, movement, transport, and warfare. When nature has not provided the ideal substance, they have bent their genius to inventing artificial materials with the properties suitable for specialized requirements. Some of these materials are meant to last a long time; others are useful only if they wear out in a specified time. All *must* have structure and functional integrity during their lifetime. A failure in this regard is the investigator's primary concern.

Our initial exercise of looking around the room is the key to our approach to materials investigation. To approach this logically, everything we see is either (*a*) nometallic or (*b*) metallic. Each of the structures—furniture, floors, supporting walls, or "supportive vehicles" (tools, instruments, machines)—is made up of some single material or combination of materials. They all may somehow fail, due to overload, misuse, age, or manufacturing deficiency. A safety specialist may be called in before a failure to design against it, while the investigator will be called in after a failure to determine what went wrong. In either case we want to know the limitations and dangers inherent in each instance.

While many things cause mishaps, they can be grouped broadly into personnel error, design defects, or materials failure. Materials failure results from improper selection of materials or oversights such as taking the wrong material from stock. Even with the proper material, there is the possibility of poor quality control when the material is processed or fabricated. Later, failure might result from overstress, fatigue, corrosion, or overexposure to high temperatures.

LOAD-CARRYING ABILITY

In all instances the material structure is under some degree of load, either from its own weight or because it supports something else. This dead weight is called a static load because it is constant. In cases where the load is intermittent (where the material is discontinuously absorbing energy), our concern is with how fast the load is applied. If the rate of loading is between one and three times the natural vibration period of the structure, then the load is called dynamic and is twice as stressful, pound for pound, as a static load. When the load onset is more than three times as fast as the natural periodic vibration, the applied load is even greater and depends on the speed of application. The force of the energy applied, here called impact loading, is now proportional to the square of the velocity of loading. For example, a 1-lb hammer lying on a bench is exerting a 1-lb load. If you drop it at a velocity of 3 ft/sec and hit the dresser, the hammer will, at the moment of impact, exert a load of 9 lb [$(3 \text{ ft/sec})^2$]. Thus, a structure adequate for a given weight may fail simply due to a faster-than-expected loading of that same weight. (See also Factor of Safety near the end of this chapter.)

Another consideration is that a structure correctly built and supported on the earth may actually fail through no fault of its own if the earth below it suddenly gives way. Here, members designed for a certain load may be called on to carry loading normally supported by other members whose integrity is destroyed when their support is snatched away. Failure in these cases does not imply a defect in the materials (although this might be involved).

In looking at the spectrum of materials that make up our world, let us set some ground rules. We are not concerned with gaseous or liquid substances, however dangerous they may be to our survival. We are concerned with anything that can carry a load, act as a barrier, transfer energy, or in short, be used as a solid engineering material. Except for synthetic plastics, which constitute a special case, the materials discussed have a long history of service to mankind, and the techniques of handling them have been well standardized. The logical place to start is with the nonmetals, and the most common nonmetal in constructions is wood.

WOODS

There are perhaps three dozen kinds of woods of economic significance, and these fall into two general classes: hardwoods and softwoods. Strangely enough, the actual hardness or softness is not the determining factor, but the cellular structure. Hardwoods come from broadleaved and deciduous trees and have large component cells. The softwoods are coniferous and have needle-like leaves and have cells. Because the softwoods grow at a fast rate and are widespread in location, they are the wood of choice for most construction purposes. Hardwoods, being more expensive and scarcer, are used largely for finishing and furnishing. All woods share the same general characteristics of growth rings. That is, the diameter of the tree increases by adding layers of wood throughout the year. The rings you see when looking at a crosscut are actually each composed of two layers. There is a wide layer called spring wood, which is laid down during periods of spring growth when

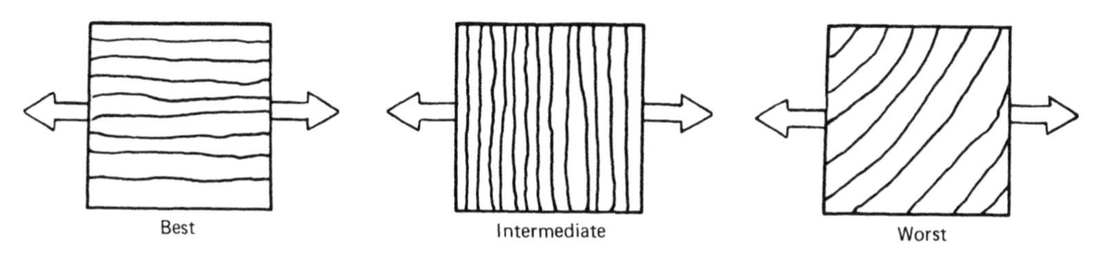

Figure 6.1 Relation of structural strength and the ring pattern.

the sap is running, and a narrow ring called summer wood, which is produced when the growth trails off.

Because boards and beams are sawed lengthwise out of the tree, the end surfaces will evidence certain characteristics of these rings. Their orientation is important for strength. Figure 6.1 shows the relationship of structural strength and the position of the ring pattern. Thus for any given load the 45° slant of the ring segments will be inferior to either horizontal or vertical configuration. This in itself may be a clue to failure.

Wood has a peculiar structure, in that the fibers are intertwined with each other in sometimes complicated patterns. This gives a certain degree of toughness, or ability to absorb energy without rupture. It is probably because of this that wood, unlike brick, stone, or concrete, shows a tendency to "creep," that is, to deform slowly over a period of time under a static load. This is a forgivable quality, since it raises a danger flag that can be detected some time before catastrophic failure, making it possible to reinforce, replace, or shore up the structure before it fails. Like most mishap situations, wood failures are the result of a combination of causes instead of a single event. For example, wood under load and partially deformed may fail if exposed to an unexpected dynamic or impact load that would normally be withstood.

The most likely causes of failure of wood structures are decay, rot, and insect infestation. It is interesting that wood kept completely dry or completely immersed in water (and not infested with boring creatures) is practically immortal and will not rot. However, this ambiance can rarely be guaranteed, and the alternate wetting and drying of exposed wood offers the opportunity for a large array of insect and bacterial activities.

To begin with, the entire United States has some degree of termite infestation. The underground variety builds clay tunnels from the soil up to the wooden support members of a building and subsequently eats away the wooden interiors. Their counterpart, and dry wood termite, can exist in the wood without needing to come out during its lifetime. Between the two, both wet and dry areas of the country are at hazard. Protection against the underground variety is assured by installing a "termite shield" at the start of construction. This is a simple but effective metal plate between the footings and floor plate that protrudes 2 inches and slants downward at a 45° angle. Such a barrier prevents the termites from building their clay tunnels into the structure. Exposed wood should be thoroughly painted or varnished to minimize the entrance of flying termites during their fall migrations. Their practice is to fly until they contact exposed wood. Then they bore into it and lay their eggs. The presence of termites is usually detected by noticing their pinhole-sized entrance holes and by deposits of brownish dust under their areas of operation.

Various bacteria are the cause of rotting. Because of this, decay-resistant types of woods are listed in Table 6.1. All trees of interest to us are composed of two types of woods: heartwood, the denser, compact wood near the center of the tree, and sapwood, the faster growing lighter wood near the outside. Heartwood is generally more resistant to decay than sapwood, because the latter contains more nutrients due to its mission of nourishing the living tree.

EARTH

The earth on which we build, travel, and conduct our business is the most fundamental of material concerns. As expanding industrial and technical society is constantly on the move, rearranging the face of the earth for factories and dwellings. Almost overnight, mushroom growths of housing tracts, shopping centers, and factories appear, the land beneath having been bulldozed and carved into convenient shapes. Trees and plants are removed, and despite observance of building codes, the laws of mechanics create a constant hazard to newly located buildings. Depending on the geographical location, soils have remarkable characteristics for supporting structures.

In hilly locations the temptation is great to terrace homes for increased population density and more efficient use of land. Two ever-present dangers exist here: rainfall and underground water. Unless properly compacted and drained—an expensive and painstaking operation—the land under any kind of structure built on a hill is subject to slipping and slumping, with attendant threat to other structures in the path of movement. Figure 6.2 shows a type of land failure familiar to Californians, the result of either or both of the water hazards mentioned above. The mishap investigator will appreciate that even an impeccably constructed building is doomed under the circumstances. In earthquake country, a similar danger exists with respect to loss of support, regardless of the care taken during construction. Little more need be said about this, since structural failures are quite apparent and may have little to do with the integrity of the building materials.

STONE, BRICK, AND GLASS

Brick and stone share a common characteristic: They are good in compression and poor in tension or shear. That is to say, brick and stone cope well

Table 6.1 Relative Decay Resistance of Various Woods

Class	Durability
I Heartwood of High Durability Even When Used Under Conditions that Favor Decay	
Cedar, Alaska	Cypress, southern
Cedar, eastern red	Locust, black
Cedar, northern white	Osage-orange
Cedar, Port Oxford	Redwood
Cedar, southern white	Walnut, black
Cedar, western red	Yew, Pacific
Chestnut	
II Heartwood of Intermediate Durability but Nearly as Durable as Some of the Species Named in the High-Durability Group	
Douglas fir	Oak, white
Honey locust	Pine, southern yellow (dense)
III Heartwood of Intermediate Durability	
Fir, Douglas (unselected)	Pine, southern yellow (unselected)
Gum, red	Tamarack
Larch, western	
IV Heartwood Between the Intermediate and Nondurable Group	
Ash, commercial white	Maple, sugar
Beech	Oak, red
Birch, sweet	Spruce, black
Birch, yellow	Spruce, Englemann
Hemlock, eastern	Spruce, red
Hemlock, western	Spruce, Sitka
Hickory	Spruce, white
V Heartwood Low in Durability When Used Under Conditions Favoring Decay	
Aspen	Fir, commercial white
Basswood	Willow, black
Cottonwood	

with heavy static loads when used properly, but they are subject to damage by impact, twisting, or shaking. Stone is becoming steadily less important as a building material but still plays a useful role in the makeup of retaining walls. It may be helpful to discuss this role.

Miscellaneous sizes of stones simply thrown together without any masonry bond are called "riprap" and are used for retaining walls in water courses or beachfronts. Coursed and uncoursed rubble, are, respectively,

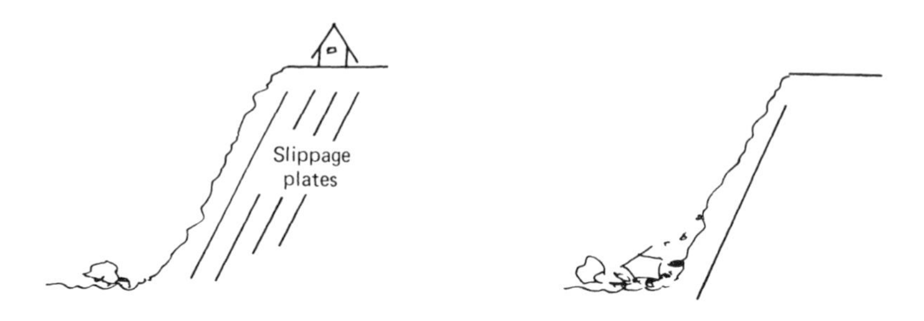

Figure 6.2 Typical California land failure.

fitted and random arrangements of rock held together by a mortar bond. Riprap and rubble are used primarily as area designators or separators and are not intended to carry static loads. Failure of these arrangements is possible, however, it generally involves excessive sideways pressures, usually the result of natural forces.

Squared stone masonry structures of exceptional strengths, built without mortar, still stand centuries after their completion. However, most brick and squared stone structures today owe their structural integrity not to their own native material but to the mortar that binds them together. Thus, a failure of the adhesive junction, instead of the units themselves, is the greatest structural threat. A true brick structure consists of two or more courses (walls) of bricks held together by reinforcing bars immersed in a bed of grout (a solidified mixture of mortar that binds the courses together and is applied while soupy). The two (or several) courses are separated by a 1-in space called the "grout space," and the combination of adhesive grout and steel bars in this space greatly strengthens the whole structure. In modern mass-produced housing, brick is almost always used as a veneer—a one-brick facing held to supporting sheathing by metal anchors. This kind of application is, therefore, cosmetic instead of structural and shares the generic weakness of all bricks and stone, that is, little resistance to tension, torsion, or shear.

For some 3000 years, glass, a supercooled liquid made by melting silica, sand, and soda, has been used both for decoration and for structural elements. Today it can be made with specialized optical properties, with a high degree of heat resistance, with conductive or insulative properties, and with a myriad of other qualities suitable for various applications.

In considering glass in its structural role as a window glazing, structural block, partitions, and so on, the mishap investigator's concern probably will focus on potential injuries resulting from impact failures. Lacerations from flying shards of window glass are still a present and common danger. Progress in controlling shatter has been made in vehicle windows and windshields, but glass windows in households and buildings still present a hazard. Shatterproof glass is made by laminating a sheet of plastic between two clear panes of glass, a process that prevents the scattering of fragments.

Investigators, however, should be alert to chemical or thermal environments that degrade the plastic film. In theory the manufacturer assures the customer that safety considerations have been met at the time of installation, but occasional manufacturing and environmental oversights may occur.

One peculiarity of glass is worth noting. Its integrity is heavily dependent on a scratchless surface. The incredible strength of glass fibers used in reinforcing resins is due to their almost perfect surface quality. A convincing demonstration of this is seen with "Prince Rupert's drop," formed by dropping molten glass into cold water. The resultant pear-shaped, long-tailed drop, whose extremely smooth surface contains a tremendous amount of compressed energy, can withstand the impact of a heavy hammer blow without failing. However, if the fragile tail is broken off, the drop will explode. This energy-containing smooth surface is used in the manufacture of tempered glass, wherein cold air blasts induce compressive stresses for greatly improved strength. There are also chemically treated glasses that can be repeatedly flexed without failure. A check of glass failure should include a review of whether the glass in question was chosen properly for physical qualities required in the use to which it was put.

MORTAR, CEMENT, AND CONCRETE

An astronomical quantity of cement products has been used in this country already, and our demand seems to increase steadily with passing years. Cement, which is actually a highly versatile liquid stone, has been a favorite building material from the days of the engineers of Rome to the present. Originally made from natural materials (Pozzolanic rock and lava), it is now a carefully controlled and reliable substance and is used in highways, dams, buildings, and for artistic applications of a bewildering variety. Since it is a rock of sorts, the material is primarily good in compression. Some ingenious methods of reinforcing it, however, ranging from the use of wire meshes and steel bars to the imbedment of glass fibers, have rendered it usable under conditions of considerable tension, flexing, and even impact loads. A British group has recently found a way to make cement more flexible, comparable to wood, by adding organic compounds based on cellulose and using the least possible amount of water.[3]

Because of its cost to manufacture, cement is rarely used alone for structural purposes, but it constitutes the parent material for various mixes of mortar (cement, sand, and lime) and concrete (cement, sand, and graded aggregate). Both mortar and concrete require water for preparation. This is a possible cause of failure, of interest to the investigator. Cement hardening is a chemical process whereby water enters into combination with the material to form an insoluble solid. Calcium carbonate and the integrity of the latter substance depends on clear, potable water. Suspended dirt particles, dissolved chemicals, and salts act immediately to degrade the product's structural integrity. Near the seashore there is a temptation to save costs by using beach sands, but this should never be done, because the presence of salt, soapy materials, and some degree of organic residues all contribute to failure. Water, then, emerges as the primary point to investigate.

The theory behind the use of filler materials (sand, gravel, and stones) is a valid and cost-reducing concept. Imagine a barrel filled with grapefruit. When the barrel is full, considerable unused spaces (voids) exist in the barrel, and one might, for instance, pour a quantity of peanuts into those voids to produce a more solid structure. Finally, the voids among the peanuts might be even more reduced by pouring in sand. The result would be a more solidly filled barrel. If we could now somehow ensure a tight bond among the grapefruit, peanuts, and sand, we could arrive at a stable, monolithic structure. This is the basis for modern concrete construction practice; a solid and uniformly strong volume of material is built by thoroughly coating strong, inert inclusions of sand, gravel, and rock with a sticky cementlike binder that securely holds all elements together. If correctly done, the result is a structure as strong as pure cement and considerably cheaper. Here again, the investigator sees a chance for possible oversight: the mixture may be poorly graded, and the water may be impure or used in the wrong proportions (this is a sensitive aspect).

As formerly discussed, mortar used for joining bricks and stones in structures employs cement, sand, lime, and water. Cement and sand could be used alone, but most building codes require the addition of lime. The reason for this is that, over time, shrinkage cracks will develop in mortar joints, and the addition of lime acts as an internal healing agent, preventing separation of the ingredients. Mortar should be applied in as soft and soupy a consistency as possible, still retaining the ability to adhere to the brick being coursed. Two factors for investigation emerge here. Fresh mortar will chemically harden in about 2 hours, so only one hour's working time should be attempted at a time. Frequently the mason will find it necessary to "temper" the mortar by adding more water, but he must guard against mixing into the batch any of the hardened mortar that may remain on the sides of the mortarboard. The second factor is the integrity of the joint itself. Most building codes require that the succeeding bricks be pushed into place, not simply laid on the mortar bed. This is required to force mortar into the rough interfaces of the brick. With reinforced, grouted brickwork it is possible to block up the grout space (the 1 in space between brick courses) by slopping too much mortar on the bricks. This results in damaging voids and perhaps interference in setting up the reinforcing bars. Such a possibility should be checked by the investigator.

Concrete (literally molded cement), like other stones and brick, exhibits brittle fracture upon failure. When laid in flat structures, such as runways and highways, integrity can be compromised by the action of the weather, most notably alternate freezing and thawing, which tears apart the internal structure. New advances in air-entrained concretes, which have a porous internal structure that allows water expansion without putting pressure on the concrete itself, do much to answer this. Another weather consideration is the exposure to sulfates from land runoff and air pollution, and special kinds of sulfate-resistant concretes combat this threat. Failure under unremarkable dynamic loadings (e.g., heavy trucks or aircraft landings) points to design deficiencies or incorrect safety factors rather than to defects in workmanship.

ROPES, CHAINS, AND SLINGS

A surprising amount of the world's work is done with ropes and chains. Failure of these two elements always occurs in tension, since this is the only operating mode. The investigator has a wide spectrum of causes to look for in this area, partly because of the materials themselves and partly because of working practices.

Ropes are made of fibers, natural or synthetic, or of metal strands and a fiber interior. Both types have disadvantages and limitations that are constantly matched against cost. Natural fibers (manila, jute, hemp, cotton, and the like) are vegetable in origin and hence are subject to deterioration from acids, alkalis, weather, and insects in addition to abrasion and cuts. Metal ropes are vulnerable to abrasion, corrosion, kinking, and deterioration from mishandling. Both are in danger from impact loads.

Manila rope, with a distinguished history of service at sea, is probably the favorite natural fiber rope. If cost versus service is an important factor, polyester rope is probably the best choice. It is stronger than manila and is resistant to insects, bacteria, and many chemicals. It is, however, seriously degraded by a solvent, benzoic acid. Nylon, another synthetic, is 2½ times as strong as manila but tends to stretch greatly, thereby storing immense amounts of energy that are released on failure, often with tragic results to bystanders and operators at the site.

Chains, usually selected for longer and harder service than ropes, have their own set of problems. Like ropes, they are vulnerable to impact loads. Even if they do not fail on the spot, they can be seriously weakened or fatigued for what would otherwise be normal service. Particular concern should be given to nicks and abrasions on the surfaces of individual links, since these blemishes act as stress concentrators that can cause disaster. The investigator faced with the involvement of a failed chain in a mishap can often learn a great deal from inspection of the remaining links. Corrosion pits, nicks, and sprung links all attest to a basic disregard of safety practices. Even with perfect maintenance, inspection, lubrication, cleaning, and storage, a chain (or a rope) can fail through improper attachment to a load. Since slings are a common method of using both chains and ropes, it is important to review a basic relationship of load to attachment. Figure 6.3 shows an often-overlooked situation and indicates that parts of a sling (in

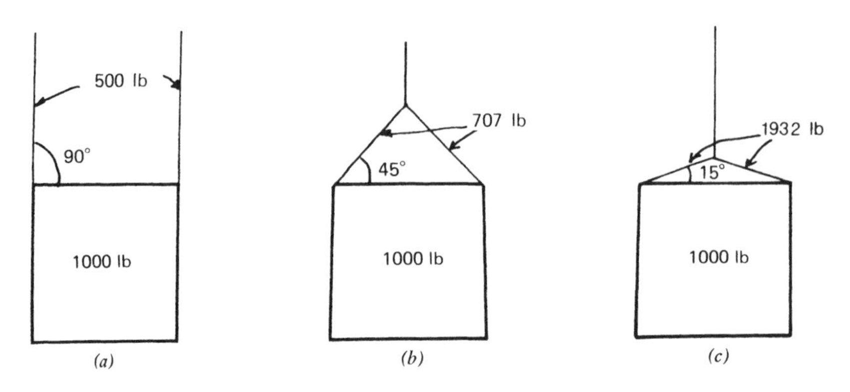

Figure 6.3 Stress on sling loads as related to induced angle.

this case the supporting legs) can actually be subjected to stress that greatly exceeds the loads they are supporting. Thus a 1000-lb weight supported by a rope or chain with a 90° angle between the sling leg and the load surface equally divides the load into two 500-lb components (Figure 6.3*a*). A cinching down of the sling steadily increases the force acting on the sling legs (Figure 6.3*b*), which can even exceed the weight being lifted (Figure 6.3*c*). The improper use of pulleys can increase, even double, the 1000-lb load shown in the figures. The resultant force varies with the cosecant of the included angle, as shown in the figure. Any chain or rope sling failure should immediately be checked for this magnification factor.

Another hazard with artificial fiber (plastic) ropes is the storage of energy during tension. In a failure, the loose ends may flail sideways with dangerous amounts of force, injuring bystanders or damaging equipment. Nylon, the chief offender in energy storage (2½ times that of manila), also presents a unique problem in generating static electricity. Obviously nylon rope would be an unfortunate choice for use with pulleys or sheaves where flammable dusts or hydrocarbon vapors are present.

PLASTICS

Artificial plastics, first appearing with the introduction of Celluloid in 1862, have undergone amazing growth and proliferation up to our present day. They are now commonplace in containers, wrapping materials, gears, toys, implements, and a host of other applications, including light structural members. When used as the latter, creep rupture and failure is a common deficiency. This is because plastics do not have any defined yield point (after which objectionable, permanent deformations take place), are generally poor in tension and shear (although good in bending), and have a long-chain polymer structure that slowly deforms under load. Where impacts are expected, special impact-resisting varieties must be used. Designers are somewhat hampered in that the usual materials testing methods (tension and compression, Charpy, and Izod) do not yield usable information. Creep tests, the only feasible approach, must not be extrapolated more than a decade (that is, 100-hr results can only be extrapolated to 1000 hr, and so on). In a fire, plastics present the added danger of toxicity of ignition products specific to the type of plastic.

METALLIC ELEMENTS

For several thousand years humans have had a continuing romance with the metals of the world. This is partly due to their strength, beauty, and yeoman service they yield in various endeavors. It is also partly because metals are curiously like man himself, behaving in a way that no other engineering material behaves. The parallel can be seen when we consider that (*a*) genetic background (parent ore from which the metal is drawn) influences characteristics; (*b*) birth defects are sometimes encountered (manufacturing defects) that may surface immediately or after years of service; (*c*) mixture with others of our kind influences our behavior (alloys show shared characteristics of physical properties); (*d*) various elements may promote disease (rust and corrosion); (*e*) too much responsibility can cause collapse (overload can produce failure); (*f*) too much repetitious activity can be exhausting (fatigue failure); (*g*) rest often helps the situation (spontaneous return to original configuration by removing stress); and (*h*) recent metals, like people, have exhibited trainability and some degree of memory.

Because metal configurations may imprison large amounts of energy, and because they are often used in dynamic systems that govern and transfer energy from one source to another, the investigator may spend a great amount of effort in examining metal failures. The particular punishment that has led to failure can be analyzed in terms of tell-tale signs we shall consider later.

First you should know something about the structure of metal as a substance. Disregarding mercury, which is a liquid, metals as a class are made up of aggregations of crystals held together by physical forces. If we are dealing with an alloy, these crystals may consist of one or more ingredients, appearing separately, or they may be composed of a mixture of the several alloying elements. As with the brick wall held together by mortar, the crystals are held together by grain boundaries, which, like the mortar, are considerably weaker than the crystals themselves. Therefore, failure is likely to occur in an encounter with a force that in some way damages or breaks the crystal boundaries. Within range of the normal variation of our tropical and temperate zones, the atoms that make up the crystals, being small energy packets in their own right, lend some important and peculiar physical properties to each metal we use. We shall consider these here.

Tensile Strength

If you firmly fasten a rod at one end and apply a stretching force, the rod will elongate as in pulling taffy. The degree to which the rod resists this stretching is its tensile strength, which usually amounts to thousands of pounds per square inch. A "weak" metal like tin or aluminum will stretch and eventually break with much less applied force than a "stronger" metal such as steel. Since this is a common cause of failure, you need to identify both the material you are concerned with and the stretching force to which it has been subjected. To a certain point (the break strength), a *ductile* metal will exhibit elongation and a reduction in cross-sectional area. Conversely, a *brittle* metal, such as cast iron, will attempt to resist the stretching force until failure, without much elongation. These two attributes give you some information on why the part failed. The surface of a brittle failure will be uniformly rough, craggy, and irregular. By contrast, a tensile failure will show two things: a brittle failure region in the center of the piece and a stretched-out region at the surface. The reason for this is simply that nothing can break until its surface breaks; the center part gives up the ghost first, while the surface attempts to hang on a little while longer. In fact, the idea of surface rupture is directly applicable to the failure of chains, for instance; thin nicks and cracks are extremely injurious. Current practice calls for the careful filing of nicks in chain links before more use, and links of a failed chain should always be inspected as a clue to the quality of general maintenance. The danger induced by a nick is not immediately obvious, but it arises from the mechanism whereby the stress level in the damaged section is amplified by the nick's presence, forcing the internal stress paths to veer sharply around the nick and dangerously concentrate them in a reduced section of the material. Thus, *any* surface modification, such as a sharp-edged hole, a sharp fillet, or an abrupt change of surface acts as a "stress riser" and can be expected to cause trouble eventually. Table 6.2 will give you some idea of the enormous variations in tensile strengths of common materials.

Compressive Strength

You can visualize this properly by taking a cubic inch of a material, placing it on a firm foundation, and adding weight until the material shows signs of collapsing. This is the value of the compressive strength in pounds per square inch.

Table 6.2 Comparative Tensile Strengths

Material	$(\text{lb/in}^2)^3$
Gray cast iron	20
Malleable cast iron	50
SAE 1040 steel, rolled	60
SAE 2340 nickel steel, oil-quenched, drawn at	
1200°F	112
400°F	282
Copper, annealed	32
cold drawn	56
Nickel, hot rolled, annealed	69
rolled, hard drawn	110
Aluminum, cast	19
rolled annealed	15
Magnesium, cast	13
rolled, annealed	15
Polystyrene	5.5–7
Celluloid	4–10
Vinyl resin	8–10
Acrylic resin	4–6

Hardness

Hardness is measured by the metal's resistance to a steel ball pressed into it or by how high it bounces a tiny diamond-pointed hammer dropped on it. In the former instance, two common tests are in use, the Brinell hardness test, which measures the diameter of the impression left by a hardened ball, and the Rockwell hardness test, which measures the depth of the indentation of a small hardened ball. From these values the relative hardness of various metals may be scaled. Specifications for machine parts and structures will usually be called out in hardness terms, and a failure might involve a deficiency in this attribute.

Impact Strength

Any force acting on a structure is greatly increased if the time of application is very short. Thus, while the structure or parts may be able to withstand a static load without failing, that same load quickly applied may cause instant collapse or rupture. This is called an impact loading and is gauged by two popular tests: the Charpy and the Izod. These tests are nearly the same, differing mainly in the way in which the test specimen is mounted in the machine. Briefly, a small test specimen of metal is clamped into a vise. A heavy pendulum is then swung against a protruding end with enough force to break off the piece. In hitting the specimen, the pendulum is slowed down somewhat, having absorbed the energy of the blow. The decreased amplitude of the pendulum, whose weight is known, then allows the experimenter to calculate the rupturing forces on the specimen.

This leads to the concept of toughness.

Toughness

Toughness is not an easy quantity to measure, but it involves the total amount of energy a material can absorb without rupture. We can intuitively imagine that hickory is tough, whereas window glass is not. In the same sense, steel is said to be much tougher than, say, cast iron, because the latter cannot absorb much impact without failing. Parenthetically, we should note that some of the new glasses are indeed so tough that they can be pounded on with a sledgehammer without breaking. Toughness is vital where repeated punishment occurs: in railroad rails (and ties), for instance.

An extensive treatment of metals would list other properties that are of interest to the investigator and, consequently we need not go more into these particular aspects for our purposes here. Consult any basic metallurgy or materials text for a more in-depth treatment.

Recognition of Metal Failure

Probably our main concern as investigators is the diagnosis of why and how metal failures occur. It might be well to keep in mind that metals and people tend to fail under similar conditions. For instance, a part or structure may be adequately made or installed to do its job well for a long time. Suddenly, through no fault of its own, some boundary condition may change (another part may fail or a fastening may come loose), and the part is called on to do more than was ever intended. Several kinds of failure may result. Table 6.3 is a guide to recognition of the type of failure.

Factors to Suspect

Table 6.3 is an abbreviated table that gives only a few of the more common hallmarks of failure, and it may be helpful to review some of the most important external influences. Experience has shown that metallurgical errors are rare. Most troubles start with improper installation, maintenance, environmental conditions, or loading difficulties. Some contributory factors are listed below.

1 **Geometry of the Part.** Notches, nicks, and other sharp changes of the surface; angular fillets; and violent changes in the structure's contours can all act to concentrate internal forces and promote internal rupture.

Table 6.3 Earmarks of Metal Fatigue

Type of Failure	Indications of Fatigue
Compressive	Structure is bulged under load. If tubular, the surface will have triangular dimples where the metal has crumpled. Usually the part wil be long and thin or will have thin walls.
Tensile	Part will be necked down at point of break, total length will be greater than original specification. Extended lips may show its break surfaces, where surface has resisted tearing apart. With brittle metal (e.g., cast iron), no elongation is noted, but surface is rough and craggy.
Creep	More common in wood and plastic than in metal. Structure plastically deforms over time. With plastics and metals, temperature is usually a factor. Loads are sometimes even moderate. Caused by slippage planes letting go gradually. Structure is bent or elongated, with or without rupture.
Torsion/shear	Structure has been subjected to a twisting stress (shaft, columns, tubes). In ductile materials, surface will show crinkling if structure is hollow, "swirling" if solid. Brittle materials exhibit a crystalline, craggy break surface.
Fatigue	Structure has been subjected to repetitive loadings that cause the planes to let go (bearing misalignment, bending shaft, flexing back and forth, etc). Break surfaces will show velvety, shiny appearance where planes have slipped, a rough surface where the instantaneous break occurred.
Crack propagation	Contributes to fatigue failure by dumping intense energy at crack tip into metal microstructure, causing crystalline slippage. Triangular chevron markings on break surface will usually point to initiation site on the break surface.
Corrosion	Surface in region of failure will be pitted, rusted, or otherwise corroded. These should be prominent in the region of the break and indicate both chemical and mechanical entrance of disruptive energy (a stress riser).

2 **Loading Speed.** Since any applied force can be enormously increased by rapid onset, a safe-magnitude static load can cause failure if rapidly applied. Even a tensile member can show brittle failure under load speeds of 50–60 ft/sec. Thus, under tension, a normal ductile necking down at the break will show materially less necking and a larger brittle surface.

3 **Temperature.** With most materials, slip resistance is increased by high temperatures. Very low temperatures can produce a brittle reaction (rubber dipped in liquid nitrogen, for instance, will shatter under a blow), while very high temperatures may result in plastic flow under loads considerably less than the conventional yield point.

4 **Surface Treatments.** Surfaces are sometimes treated to produce a more resistant structure. Certain of these treatments introduce a compressive stress at the surface, thereby lessening tensile strength from an applied load. Such treatments include shot peening, carburizing, nitriding, and so on. What sometimes happens, however, is that the surface itself will develop a flaw or crack that will propagate into the structure.

5 **Contour of the Structure.** The ideal structure has gently sloping and rounded approaches from one surface to another. Any violation of this can cause trouble, as with flat-bottomed holes, keyways, slots, and sharp fillets, where the sharp change of direction acts as a stress riser. Cracks tend to start here.

6 Residual Stresses. Some parts, particularly forgings, may have stresses left in the metal at the time of fabrication. These may act in concert with externally applied loads to cause failure. This cannot be detected per se, but a manufacturer's history may shed some light on this.

7 **Resonant Vibration.** Applied forces that coincide with the natural period of vibration of the structure or configuration are amplified, sometimes dangerously. Thus, steel building structures may collapse from earth shocks that hit their critical period. Machine shafts and moving parts may be similarly damaged due to loose bearings, incorrect mountings, and other kinds of service wear.

8 **Torsional Fracture.** Torsional fractures have several characteristic appearances. Tensile failures often occur on a 45° inclination to the axis of the piece; torsional fatigue fractures often show as a rubbed-smooth surface around the periphery, and others show a star-shaped fracture. Again, the cause may not be inherent in the piece itself, but can happen as in an overload, or when a bearing binds, throwing all the driving force on the shaft structure instead of passing it to other structures as designed.

NEW MATERIALS

In the quest to develop better materials there are always tradeoffs. As new techniques are developed to increase strength, it becomes more difficult to maintain or improve properties such as ductility and resistance to fatigue, cracks, stress corrosion, and extended exposure to high temperatures. The improved properties can improve performance, increase reliability and safety margins, or provide a combination of these attributes. Sometimes the pressures of using these gains solely for improved performance may lead to more mishaps. The demands for greater performance may force us to sacrifice safety margins, reliability, and maintainability. The premature use and subsequent failure of these new materials is more a management failure than a materials failure. The typical development cycle for new materials may be 10 years or more depending on the complexity of the technology involved.[4]

The latest aluminum alloy (mostly used for aircraft) contains 2–3 percent lithium. This increases the elasticity or stiffness by about 10 percent. Its high strength allows weight reductions of 10 percent over the more traditional zinc alloys. However, there are some tradeoffs with toughness and resistance to stress corrosion.

For composites, the use of graphite epoxy will continue to expand. Improved graphite fibers will yield more structurally efficient composites. To increase the reliability of aircraft composites, several improvements are being developed. First, the addition of 10 percent rubber will improve toughness and resistance to brittle fracture. Ceramics continue to be a good possibility but tend to have brittle failures at high temperature.

FACTOR OF SAFETY

One of the most comfortable approaches used in engineering designs is building in the factor of safety. (This is also called the factor of ignorance in some circles, for its function is to provide a margin of error that allows for a considerable number of corollaries to Murphy's law without threatening the success of an engineering endeavor.[5])

The factor of safety calculates the load that causes failure by the maximum load expected to act on a structure. Thus, if a rope with a capacity to lift 6000 lb is used in a hoist to lift no more than 1000 lb at a time, then the factor of safety is 6000/1000 = 6. We can expect operators to routinely overload it by half or 1500 pounds, reducing the safety factor to 4. Operating the hoist in a jerky manner could double the load to 3000 lb, resulting in a safety factor of 2. An inferior rope could easily deteriorate to where it could barely support 3000 lb. This could conceivably result in a safety factor of 1. Thus a hoist routinely operated in a jerky manner could exceed the safety factor we now have—resulting in a failure. A safety factor of 6 was not good enough to foresee all possible conditions.

STRUCTURAL FAILURE

Structural failure refers here mostly to building structures but can occur in any type of structure. A general discussion of the subject is included to give the lay investigator, who is not a structural engineer, an overview of the subject.

Since excess strength can be unattractive, uneconomical, and unnecessary, engineers must make decisions about how strong is strong enough by considering architectural, financial, and political factors as well as structural ones.

When a structural failure occurs, a large factor of safety is built into subsequent structures. When groups of structures become familiar and suffer no unexplained failures, there is a tendency to believe they are overdesigned, that they have an unwarranted high level of safety. Confidence in the structure mounts and engineers develop a high factor of ignorance with respect to structures they are so familiar with.

Some Causes of Structural Failure

In the following list the human element seems separated from the physical environment in which the structure is placed. How well the designer considers the categories and incorporates design features to guard against structural failure from these elements will determine the safety of the design.

This is, necessarily, not a complete list. You could probably think of some additions, and investigators will not always agree on the category in which a particular item should be placed, but it will serve our purposes.[6]

Limit States

Overload: geophysical, dead load, wind, earthquake, etc.; man-made, imposed, etc.
Understrength: structure, materials, instability
Movement: foundation, settlement, creep, shrinkage, etc.
Deterioration: cracking, fatigue, corrosion, erosion, etc.
Random hazards
Fire
Floods
Earthquakes
Explosions: accidental, sabotage
Vehicle impact

Human-Based Errors

Design error: mistake, misunderstanding of structural behavior
Construction error: mistake, bad practice, poor communications

A subcommittee of the U.S. House of Representatives Committee on Science and Technology held hearings in 1982 on the problem of structural failures in the United States. They found six "critical" areas that allow structural failures to occur. They could also guide the investigator:

1. Communication and organization in the construction industry.
2. Inspection of construction by the structural engineer.

3 General quality of design.
4 Structural connection of design details and shop drawings.
5 Selection of architects and engineers.
6 Timely dissemination of technical data.

Also mentioned were cost cutting on design and construction and, to a lesser extent, adequacy of codes and the impact of "fast track" scheduling.

To keep designers aware of ways in which the type of structure being designed has failed in the past, an archive of structural failures has been funded by the National Science Foundation. The civil engineering–oriented Architecture and Engineering Performance Information Center (AEPIC) is located at the University of Maryland. An engineer can call up a wealth of relevant experience on a design or material he is using. It would seem to be an excellent resource for the investigator.

CRACKS

For several decades we have viewed the fatigue process as having two stages. Microscopic cracks develop at "nucleation sites"—points of material weakness or stress concentration—during the first stage, which may last as long as one-half the entire lifetime of a machine part or structure. As repeated loading continues, these cracks grow, and some combine into a macroscopic fatigue crack. During the second stage, this crack grows at an accelerating rate as the loading cycles continue. If the crack reaches an intolerable size for the load being applied, the weakened structure can no longer support the load and breaks. The crack makes a final advance under a load level that may well be within the design load of the unfatigued structure. Fifty to ninety percent of all structural failures result from crack growth. Usually the cracks grow slowly. They are not always a cause for alarm but engineers should take them into consideration.

Fracture mechanics is an analytical tool used to predict the behavior of materials in the presence of cracks. The field developed primarily because of the need to understand how small cracks in high strength material could drastically reduce the strength and load-carrying ability of high strength steels. Fracture mechanics uses the concept of a critical defect in a component under load and calculates the size of the defect at the time of final catastrophic crack propagation. In mishap investigation, by examining the failed part to determine the size of the defect and using fracture mechanics, one can estimate the magnitude and direction of loads that were necessary to cause the final failure.

All machine and structural failures are problems in fatigue; the forces of nature are always at work, and every object must respond in some fashion.

Theoretically, fatigue can be avoided, but overdesigning structures so that peak stresses never exceed the threshold level is not practical. Optimal design allows for cracks, but ones that progress so slowly that the product can be taken out of service before they pose a problem.

Quality control should eliminate unacceptably large flaws by minimizing deviations from an acceptable norm and rejecting inferior workmanship. Unfortunately, the techniques for detecting flaws are wanting. Not only are the instruments insensitive, but their use and interpretation are often more art than science.[7]

SUMMARY

This chapter is obviously not an exhaustive treatment of materials failures, but the highlights provide a general map in your quest to explain materials failure. It has been estimated that eight or ten good hours of instruction in materials will enable an investigator to perform creditably in the area, particularly in recognizing the type and nature of failure involved and, above all else, in knowing when to call in an expert. As materials developments progress and the number of materials and manufacturing processes and the complexity of their uses grow, it is unlikely that a few hours of instruction will serve the investigator. If nothing else, this chapter should provide the lay investigator with enough background in materials to know when to call an expert.

QUESTIONS FOR REVIEW AND DISCUSSION

1 Are bricks better in tension or in compression? Why?

2 How does creep differ from fatigue?

3 What is meant by toughness?

4 Define and explain the mechanics of a "stress riser."

5 Why are filler materials used in concrete?

6 Name four causes of deterioration in wire ropes.

7 Why is water important in concrete mixing?

8 Name two disadvantages of nylon rope.

9 State some drawbacks of natural fiber ropes.

10 Explain what is meant by "factor of safety."

11 What is AEPIC? What does it do?

QUESTIONS FOR FURTHER STUDY

1 Bring a clear-cut example of a "brittle" failure to class.

2 Bring an example of graphite fibers or a graphite fiber product to class.

3 Find an article dealing with the use of ceramic materials in a modern, high-technology application, and bring a copy of it to class. Be prepared to brief the class or your instructor on the contents.

REFERENCES

1 McKaig, Thomas K., *Building Failures: Case Studies in Construction and Design*, McGraw-Hill, New York, 1962.

2 Haddon, W. J., in *Directions in Safety*, Ted S. Ferry and D. A. Weaver, Eds., Charles C. Thomas, Springfield, IL, 1976, pp. 87–98.

3 Hansen, James, "The Delicate Architecture of Cement," *Science 82*, December 1982, pp. 49–55.

4 "New Materials Can Reduce Mishaps," *Safetyline*, Mar. 14, 1986, pp. 2–3.

5 Petroski, Henry, *To Engineer Is Human*, St. Martin's Press, New York, 1985.

6 Blockley, D. I., *The Nature of Structural Design*, Ellis Horwood, Chichester, England, 1980.

7 Osgood, Carl C., *Fatigue Design*, 2nd ed. Pergamon Press, Oxford, 1982.

OTHER REFERENCES

Brown, J. Crozier, "Anatomies of Computer Disasters," *Proceedings of the First International Conference on Computers in Civil Engineering*, American Society of Safety Engineers, New York, 1981, p. 250.

Clausee, H. R., *Industrial and Engineering Materials*, McGraw-Hill, New York, 1975.

Madayag, A. F., *Metal Fatigue: Theory and Design*, Wiley, New York, 1969.

Metals Handbook, Vol. 10, *Failure Analysis and Prevention*, 8th ed., American Society for Metals, Metals Park, OH, 1975.

Moore, H. F., *Materials for Engineering*, McGraw-Hill, New York, 1941.

Schlenicee, B. R., *Introduction to Material Sciences*, Wiley, Sidney, Australia, 1969.

CHAPTER SEVEN

Systems Investigation

The common approaches to mishap investigation can be classified as "traditionalist" and "systems" approaches. The traditionalist approach considers a sequence of events along a single track that result in a mishap. The systems approach, as the name implies, involves looking at an entire functional system or subsystem. It also delves into the interaction of systems that may or may not have an apparent relationship.

While the traditionalist seeks unsafe acts or conditions, the systems person will look at what went wrong with the system, seeking something wrong with the system operation that allowed the mishap to take place. It calls for understanding the theory of systems, be they mechanical, biological, social, etc. It also calls for looking at what went wrong with the system so we can correct the entire system, not just one part of it. In Chapter 14, for example, we will look at the management system, and throughout the book we will find ourselves looking at different systems. In this chapter we will deal with operating and hardware systems to detect the many things that can go wrong with a system and cause mishaps. We will also use the systems approach as an analytical tool that allows us to organize the investigative approach into subsystems.

A hardware operating system may be part of a larger, more sophisticated system. For example, if the larger system is a complex piece of production machinery, a locomotive, a printing press, or an aircraft, it will contain power, hydraulic, electrical, and lubricating systems.

The systems investigation seeks the conditions and capabilities of each system before the mishap. In this chapter we look at power, electrical, hydraulic, lubricating, robotic, and computerized systems.

The systems approach is typically used in large-scale investigations. A team may be used to check the hydraulic, computer, electrical, navigation, control, fire detection, fuel, or some other system. There is no reason, however, why an investigator working alone cannot use the approach to find areas of possible cause and reject from consideration systems that are not involved.

SUGGESTED APPROACH

A few suggestions[1] for conducting a systems investigation follow:

1 Before starting an investigation, obtain or develop a schematic of the system you will be investigating.
2 Review the mishap site to study general conditions; do not just tear into the site and start examining systems. Determine first what systems might be involved. Know what systems should have been working at the time. A single-auto mishap at night would certainly involve the vehicle's lighting system, but it would not be much of a factor in a good-weather daytime investigation.
3 Consider and record what did or did not happen with a particular system, that is, what was and was not working. Both positive and negative information is desired.
4 Go through a suspect system subsystem by subsystem and part by part.
5 Recover all parts of the system from the mishap site, and be alert for missing parts.
6 Do not disconnect parts of the system, unless it is essential, until the entire system is viewed and it appears necessary to go into a detailed investigation of part of the system.
7 Do not attempt to tear down or analyze part of the system that exceeds your technical capability or that of the person performing the examination for you. Judge whether the person is capable or qualified to perform that level of analysis on the system.

THE TECHNIQUE

We want to know if a particular system contributed to or caused the mishap. Did the system operate properly? Can it yield any information on the mishap circumstances? A sound approach is to assume that a malfunction did exist in the system and then proceed to find out what it was.[2] Guidelines follow:

1 Every system has signs of the expected operating conditions (switches on or off, valves turned, etc.). Look for them.
2 Maintenance records and logs can show chronic conditions or recent troubles. Look for the reasons for past complaints and corrective actions.
3 Know generally how the system works. Look for evidence of common malfunctions.

4. Determine whether any system components are missing.
5. Are the system components the correct ones; valves, circuit breakers, and so on? When examining a component, determine if it is all there. See also if too much is there (wrong parts or foreign objects).
6. Check for compliance with service bulletins and directives on the system.
7. Look for something unusual in the system's color, smell, shape, location, position, general appearance, or damage. When possible compare it with a similar unit. The obvious is sometimes hardest to find.
8. If others are working on related systems or if parts have gone to a lab, evaluate the findings of all parties.
9. Review the observations you have made on a system, and isolate the pertinent ones, either positive or negative. Is there anything in your observations that needs amplification? When several weak or inconclusive observations agree, that indication has merit and value.
10. Will the condition of one system reveal itself through indications in another system?
11. Combine the observations of all systems, and attempt to find a pattern, trend, or overall picture that will account for most of the observations.
12. If observations appear unrelated or contradictory, try putting time intervals between them. Imagine a malfunction that would allow the contrary observations to exist.
13. If your observations support a theory or idea, discuss it. Expect criticism, and realize that many times this will support your idea.
14. If your theory holds up, now is the time for bench testing to prove it.

With these guidelines in mind, we will now examine, in a general way, signs of failure in four types of systems: power, hydraulic, electrical, and lubricating.

POWER SYSTEMS

For the discussion at hand, power systems are engines of most types: reciprocating, rotary, diesel, gasoline, or turbine. Our main concern in investigating power systems is to determine whether the proper power, indeed if any power, was being provided at the time of mishap. While certain engines have components that are not present on others (fuel pumps and diesels or pistons and turbines), the principle of damage to moving parts is our interest.

Picture a table fan with plastic blades. It sits in front of you and is turned on. If you were to stick a thin peg into the blades, it might be cut off by the first or second passing blade and the fan would keep going. A larger peg might deflect and survive several blade passes before being cut off. If you turned off the fan and stuck the peg into the coasting blades, the damage to both the blades and the peg would be minimal. The blades would quickly stop, perhaps at the first contact, but there would be marks on the peg. If you thrust a metal rod firmly into the coasting blades, it would surely stop them at first contact. If the blades were still under power, they might be sheared off, the motor might be forced to stop, and there would be evidence of the motor overheating. The fan might continue to operate with the blades shortened a bit. This typifies the damage that results to an engine when power is being applied or when it is coasting at high revolutions. If the fan was stopped completely and the metal rod was pushed into one of its blades with reasonable force, the blade would be bent, marked, and perhaps broken, but damage would be limited to that blade in contact with the rod. The important fact to note is that when the engine is under power the damage will be far more extensive; there will be several indications, both externally and internally, of the presence of power in the system.

As we discuss engines and power in this volume, we are dealing with only a minute facet of the subject. A power plant failure investigation requires the services of an engine expert.

Engine Power Factors

If substantial power is being developed at the time at which the engine is suddenly stopped due to a mishap, you might observe the following:

1. External rotating parts such as propellers or fans will have extensive damage, with the blade tips curled forward.
2. A torsional shaft will be displaced: splines on the shaft of drives, including generators and accessories, will be offset, that is, twisted opposite to the direction of rotation. Realize, however, that many shafts are meant to shear under stress rather than cause overheating and a potential fire.
3. If rotating parts were under power, there will be internal engine damage with stripped gear teeth. The rotation may also have carried out a machining action, such as cutting a circle on the radiator or scoring the inside of a gearbox. Expect to see a uniform type of

damage: blades bent the same way or worn in the same manner, or nearly perfect scoring marks.

4. If an exhaust or hot section has high operating temperatures, expect a plastic bending (not a brittle breaking), possibly without breaking.
5. Inside the engine, if engine failure was suspected, look for signs of massive damage as well as signs of failure and fatigue.
6. The positions of switches, throttles, and governors provide signs of the expected power, but this is an indication only, not an assurance. Such evidence must be correlated with other evidence.

None of the factors by itself is sufficient evidence. Further, the apparent application of power at the time of mishap may not reveal malfunctions such as internal failure, a stuck throttle, or a failed governor.

The next obvious question about power is whether the engine failed and, if so, to what extent. That is, was sufficient power being developed? This is approached by asking, "Did we have an engine problem?" If the answer is yes, failures can be divided into two types:

1. Failure due to breakage.
2. Failure from a loss of power.

If the failure was due to breakage, we must pursue a fatigue/stress type of investigation. If the problem was a loss of power, there are three possibilities:

1. Engine ran smooth and then quit.
2. Engine ran rough and then lost power.
3. Engine ran smooth and then lost power.

These three failures can be investigated as either operator-induced malfunctions or maintenance-induced malfunctions.

This slight bit of organization will ensure that the investigator examines most power system failures in a logical sequence.

Signs of Little or No Power

If the mishap has been severe enough to badly damage the engine, then the absence of rotational damage or other signs of power is evidence of no power. If the engine was not developing substantial power, then the damage is restricted to impact damage.

If the mishap was not severe, we may need to look at adjoining or dependent systems to see if power was being applied. For example, most power brake systems require the engine to be running, and most engines need to be running for the electrical system to be generating electrical power (do not confuse this with battery power).

ELECTRICAL SYSTEMS

In electrical systems, as with engines, our first problem is to find out if power (electricity) was being applied at the time of the mishap. Although power may have been available, this does not mean that it was being used as required. A component may have failed even while electrical power was available to the component.

Electrical Power Factors

A detailed examination of the electrical system may not be needed if we can be certain power was received in the right amount at the right place. For example, if we are concerned about power to a radio system, evidence that tubes or lamps in the radio were on during the mishap tells us that electrical power was available. It divulges nothing about the reliability of the radio.

Electrical motors rotate at exceptionally high speeds and tends to coast for a long time (often many minutes) after power is shut off. The evidence from power-on and coasting operations is often different as discussed under power systems. Most of the operation is rotational, so we should look within the motor for rotational damage, but not before bench checking in an electrical shop for operation. If the power was off, but the motor was coasting at a high rate, the rotational damage will be much less than if the power was on. Where a mishap has caused sudden arrest of the internal rotating parts but little rotational damage, we can assume a coasting or no-power situation. If there is extensive scoring and wearing, it would be safe to assume that power was being applied. Characteristic rotational scoring on the direct current (dc) commutator of an inverter armature indicates powered rotation at impact.

Examination of broken drive shafts will show wearing and scoring at the break if power was being applied at impact. The shaft ends, for example, may be polished smooth from the wear. Lack of power would be indicated by bending and nonrotational displacement of the spline. Since most shafts are designed to fail under extreme loads, broken shafts may also indicate a problem with the driven component. If the shaft did not fail under load conditions as intended, a fire might have resulted from the generated heat. If a shaft under power has been sheared due to impact, the type and direction of shear will indicate rotational damage. It is unlikely that an impact strong enough to shear a shaft under power would also immediately halt rotation.

Shorts and Switches

If power is not being received at the suspect component, then the generating system is suspect. Proof that power was being generated places our trouble between two items: the generator and the component.

Many systems are both electrical and mechanical, that is, there is a mechanical linkage to or from the electrical source. It must be determined which failed first. Typical items to examine include switches, circuit breakers, wiring, and terminals. These should be examined for signs of arcing and shorting. Switches may be installed incorrectly, they may be undersize for the expected load, or their boxes may trap moisture and short out. At this stage we can encounter human engineering problems with poorly placed, hard-to-read, and difficult-to-reach switches. Circuit breakers are sometimes used as switches and may not function properly because they are the wrong type or size. Wiring and terminals should be checked for tightness of connections, proper installation, the presence of unauthorized jumpers, and the use of the wrong kind of wire for the load or environment.

Shorting in the electrical system often leaves no trace and thus no physical evidence.[3] Nevertheless, the search for shorts and overloads can be most rewarding. Where there has been a fire the remaining electrical system may still provide useful evidence. It is difficult to trace some circuitry, but an experienced electrician can help unravel the mystery.

Evidence of corrosion between a stud (brass or steel) and a bus (aluminum) indicates a loose connection. Check all connections of the circuit protectors attached to a bus bar and all wiring of the bus bars and terminals for looseness. Arcing between terminals may also be due to dust, lint, moisture, and wire strands, particularly when terminal strips are exposed.

If a fire has resulted from internal overload, the internal wire strands will be beaded, annealed, or darkened. Wire damage from external fires will char the insulation but leave the inside strands bright and the inside of the insulation barely damaged.

Even if the wires are melted from external heat, expect the inside strands to be bright, and the ends will not be beaded. Shorts will sometimes be signaled by burned areas where the short occurred, even to the extent of parts or bolts being melted, much as you would expect from lightning damage.

In today's complex circuitry systems it is unlikely that the investigator can find his way without the aid of an expert, even with the help of diagrams. He can, however, see if there is evidence of failure and call for the expert to assist from that point on. To repeat an often-stressed observation of this book: Knowing when to call in the expert is a characteristic of the professional investigator.

Control Devices

When a generator operates under excessive load for more than a short time, the increased heat will discolor it. Sometimes the commutator will rise and interfere with the brushes, which will later be broken off or carried away. A short circuit in the generator will also cause overvoltage and may burn out filaments and tubes.

Generator problems also include overloads due to an emergency situation or procedure and thus are only symptomatic of other problems. The true state of a generator may not have been realized: Perhaps warning lights failed and the operator had no warning of the failure of a voltage regulator or reverse-current regulator.

Light-Bulb Analysis

A mishap analysis may hinge on whether a certain light was on or off at the time of a mishap. This determination has been developed to a science, and there are experts in most countries that specialize solely in this area. A 30-power (30×) magnifying glass will serve the investigator in lamp inspection, although a laboratory technician may use far more powerful magnifications for further investigation.

The subject may best be covered by making some observations,[4] first about undamaged bulbs (lamps) and then about damaged ones.

Normal Lamps

1. New bulbs are clear. The filament is usually a small, tight, even, smooth, bright coil of tungsten wire clamped at each end into steel wire supports. A bulb may have two filaments, such as one for a high beam and one for a low beam, or for a stop and tail light. The coils will be evenly spaced, and the base should be bright.
2. An aged lamp is darkened as the tungsten evaporates from the filament and weakens it. This is most conspicuous in large bulbs. The filament may sag with age and be pitted but will likely still be bright. As the tungsten evaporates, the filament gets thinner and the concentrated current gets hotter at the weak spot, eventually burning out. The current will cause overheating and filament failure, and the filament ends will become round or beaded.

In a collision, rapid acceleration or deceleration subjects the filament to violent inertial forces. The severity of impact and localized forces applied directly to the bulb are the most important of several factors in lamp mishap analysis.

1. If the bulb is hot (light on), the filament will be stretched out of shape and will not return to its coil. It will appear as an irregular stretch as

opposed to a uniform sag. In a hot shock it is usually stretched more near its ends.

2 In a cold shock (bulb off), if the glass does not break, there will be no effect on a filament unless the shock is severe. If it is severe, a break of the filament will be shown as a brittle break, not plastic as when it is hot. The filament might even be broken loose within the bulb's glass container. The accuracy of lamp condition at time of an impact is based on the quickness with which a filament cools on the application or a power shut-off. Only a fraction of a second of power being on or off permits one to view the lamp as "hot" or "cold."

3 If the glass is broken (or even cracked) during a hot break, air will cause the filament to oxidize and disappear.

Guidelines for organizing the condition of a lamp after a mishap follow[5]:

If the glass is broken, check for:

1 Filament color.
   (*a*) Blackened.
   (*b*) Tinted.
   (*c*) Bright.
2 White oxide.
   (*a*) On glass, supports, stem.
   (*b*) On filament itself.
3 Fused glass on filament.
4 Base support burned, melted, pitted.

If the glass is either broken or unbroken, check for:

5 Filament broken or unbroken.
   (*a*) Glass etched.
   (*b*) Stretched out, uncoiled.
   (*c*) Moderately elongated.
6 Filament broken.
   (*a*) Ends fractured, angular.
   (*b*) Ends melted, tapered, round.

Checking the six items above against the following will give an excellent summary of the lamp condition at the time of mishap:

1 Filament incandescent.
2 Other filament incandescent.

3 Filament hot but not incandescent.
4 At least one filament incandescent.
5 Filament cold.
6 Filament burned out.
7 Lamp energized, power on.

The foregoing coverage of the extremes of bulb evidence should give the investigator an indication whether an expert is needed or whether there is even a problem.

Nickel-Cadmium Battery

The common lead acid battery no longer meets the load demand of many situations. The nicad (nickel-cadmium) battery is coming into common use because it can discharge and recharge quickly. Due to certain characteristics that permit this rapid discharge and charge, the nickel-cadmium battery is particularly susceptible to heat both in ambient temperatures and in generating its own heat. These batteries are most often enclosed in compartments, compounding the heat problem. To sum up the problem, a condition known as thermal runaway builds up, which results in excessive heat and explosion or fire. In the process, hydrogen gas is generated, and a spark can set off an explosion under the right circumstances.

Radar

Powerful radar installations, mobile or permanent, have the ability to injure through burns and to ignite combustibles at a considerable distance when they are in the radar beam. Check this possibility if a mishap occurs near radar beams.

Laser

The commonplace use of lasers has generated a new set of personal injury hazards as well as the capability to ignite combustibles at a distance of several miles. Lasers have many different capabilities and pose related hazards. The potency of some lasers is indicated by their consideration as a weapon of war. A technical expert is clearly needed if the possibility of a laser mishap exists. Since the beam can be damaging even when reflected or bent by weather elements, more than a line-of-sight involvement is possible.

Carbon Fibers

Some of the new composites coming into common use, including carbon fibers,[6] have created some unusual hazards. At this writing the electrical hazards associated with carbon fibers appear to be minimal, although the possibility for great damage always exists. Fibers emerging from the surface of composite materials are small, sometimes invisible to the naked eye, and can cause shorts when they bridge an electric circuit. While the possibility of a catastrophe always exists with electrical shorts, NASA,[7] in late 1979, judged that the release of the fibers in crash fires (which was the main concern) constituted an inconsequential risk.

Protection

Shock is the most obvious electrical hazard[3] and the one that hurts the most people. People can be killed or burned by the heat of an electric arc, particularly in a confined area with many grounding sources. During an electric short circuit a tremendous amount of energy is released. This can cause explosions with doors becoming missiles, the release of large quantities of oil from transformers, or the activation of power system disturbances that affect entire systems.

Loose Electrons

One expert[8] summed up this situation by explaining that there are no loose electrons running around in the wreckage and that "computer systems rarely leave physical evidence of malfunctions." That is, in investigations of electrical mishaps there is often no sign, no indication of currents gone astray. In many modern systems there is little to indicate electrical malfunction, particularly where a complex electronic system has been necessarily controlled by an even more complex and difficult-to-track computerized system.

Fortunately, techniques are continually being developed that can track the path of electrons prior to the occurrence of a mishap and interruption of power. It has been possible in some recent investigations whose importance justified a considerable resource expenditure to recreate the precise setting of controls and computer inputs at the time power was interrupted. For example, bubble memories in computers retain some features of previous activity. Literally "saved," these memories can then be reconstructed, sometimes by software programs designed for the purpose.

HYDRAULIC SYSTEMS

Hydraulic systems usually function in conjunction with one or more other types of systems. Commonly we find mechanical, electrical, and hydraulic systems working together. Knowing that a hydraulic system did not fail may point to a connecting system. Therefore, that positive input is important, just as important as knowing that the hydraulic system did fail.

General

There are two basic types of hydraulic systems: open and closed. In an open system, all pump output is directed into the reservoir when all selector valves are neutral. In this way the pumps are not loaded. In general, the closed system maintains constant pressure in the system and requires a system bypass valve to unload the pumps when no hydraulic service is required. Regardless of the type of system, most can be investigated by considering five aspects: supply, pressure, control, protection, and application.

Hydraulic systems often operate under extreme pressures such as thousands of pounds per square inch. Under such pressures a nearly invisible pinhole leak can cut a person's hand as surely as a knife and greatly influence system supply and pressure. Typically we find hydraulic systems used to control power pressures and heavy machinery, lift heavy devices, operate marine or aviation controls, and control a host of weapons.

Examination

When isolating hydraulic systems for examination, care must be taken not to lose any fluids or pressure until fluid levels and pressure can be recorded. Detailed examination of a hydraulic system should include the following:

1. Examine the general condition of the system, including condition of external lines, internal passages, and oil plugs.
2. Check servicing and maintenance records for prior problems and compliance with directives.
3. Examine the condition of hydraulic pumps, and if suspect, bench test them.
4. Heavier hydraulic fluids tend to be easily contaminated. Check plugs, the fluid itself, and filters.
5. Ideally, both damaged and undamaged systems should be checked for operation. In a severe mishap this may call for checking the system item by item, once it is available.
6. Check all fluids closely for contamination. The close tolerances in all high-pressure hydraulic systems make this essential. Handle the fluids carefully to avoid more contamination. Fully identify fluids you send to the lab; include any possible contaminants of soil, moisture, fuel, and so on, if the fluid sample was gathered outside the reservoir.
7. Because the system is often closely tied to other systems it may not be possible to check the hydraulic system by itself. For examination purposes, consider associated pumps, screens, filters, actuators, and so on as parts of the system.

8 Make certain whether there is a need to examine each component. Anything from a plain functional test of one component to a full-blown operational test of the system may be required.
9 Exercise great care in removing a system to avoid more injury or damage. Hydraulic fluid is not only under great pressure but is toxic. Particularly protect the skin and eyes.
10 When any part of a system is disconnected, plug, cap, or seal it to prevent more contamination.
11 Some hydraulic units contain powerful springs, particularly those systems with relief valves. Take care when disassembling.

LUBRICATING SYSTEMS

The failure of a lubricating system can be traced to one or more of three things: too little lubricant, too much lubricant, and/or contaminated lubricant. A shortage of lubricant will cause overheating, seizure, and system failure. Evidence of overheating is seen in discoloration and nonlubricated contacts. Too much lubricant, which is a common fault, usually involves bearings and fittings that allow grease to be added manually. When there is too much grease the bearings churn in the lubricant instead of turning, and thereby generate excess heat. Bearings overheated for this reason will have a bluish color. Older equipment commonly used all-purpose lubricants, but today's sophisticated machinery with close tolerances and high temperatures calls for specialized lubricants that are especially sensitive to contaminants.

Contaminants

Some metals and other substances in modern systems are vulnerable to erosion by certain lubricants. This makes the lubricant itself a contaminant. The investigator must determine whether the proper lubricant has been used. Many lubricant problems can be traced directly to contaminants. A contaminant can be almost anything added to the correct lubricant. It could even be a lubricant of the same specifications from another supplier. The main concern with contaminants is the addition of thinners that destroy the lubricant's integrity or contaminants that cause it to act as a grinding compound. A few grains of sand, dirt, or metal will destroy any piston, valve, or bearing. Evidence of this damage can be seen, often by the naked eye, in the scoring, scratching, and rubbing of pistons, valves, or bearings.

Chemical Contamination

This can come from accidental mixing of petroleum products, which in turn causes deterioration of seals and gaskets and clogs lubrication jets. The seals lose their ability to seal, and this results in a loss of oil and subsequent system failure. Some synthetic oils attack copper, forming harmful by-products. Use of mineral-based oils in place of synthetics may result in heat breakdown and a loss of cooling capacity. A common contaminant is fuel introduced into the lubricating system, which thins the lubricant, causing a lack of cooling capacity and sometimes fire. This is evident in higher consumption and operating temperatures.

Examination

The scope of the examination depends on the available facilities and personnel. Secure a sample that is as clean as possible and typical of the system. Laboratories can tell precisely the normal and abnormal content of sample contaminants, but the investigator must tell the lab all he can about possible sources of contamination and furnish all the details when submitting the sample. Disassembly of components should be left to the experts and in any event should be carefully documented. Providing the examiner with all possible data can shorten the lab examination by days.

Bearings

The very high rotational speed of bearings, often 40,000–60,000 rpm, makes them susceptible to lubrication-based failures. Such failures can result from sheared lubricating pump drives, clogged filters and jets, improper lubricants, damaged oil seals, contaminants, ruptured lines, and so on. The following observations can be made about bearing failures:

1. Loss of lubricant results in an overheated appearance and is the most common cause of bearing failure.
2. Insufficient cooling results from both too little and too much lubricant.
3. Fatigue failure of bearings is uncommon and not usually a lubrication failure. Most often it comes from repeated shock and improper lubrication under some circumstances.
4. Lack of lubrication can sometimes cause race skidding, which can be identified by skid marks on the balls and rollers.

Bearing failures are serious, sometimes catastrophic, and can cause the failure of any component or system that rotates via bearings. Sometimes a system will use hundreds of different types of bearings and several different types of lubrication. As in many specialized areas there are bearing specialists who can furnish exact information.

ROBOT SYSTEMS

The exact number of robots in use is not known. What with their increasing numbers (expected to be 100,000 by 1990) and the number of mishaps they have already experienced, as an investigator you will almost surely be called upon to investigate a mishap in which a robotic device is involved.

Robots are ideal examples of both simple and complex systems. They are often subsystems of a larger system such as a manufacturing cell, an automated process, or even a completely automated factory.

Every system mentioned in this chapter is common to robots. Often the robot has an electrical, hydraulic, and lubricating system and interfaces with many more.

Examples of Robot Mishaps

Suppose a large manufacturing robot has a problem such as a broken hydraulic hose, and a long assembly line is halted as a result. This is a mishap as far as we are concerned. There has been an unplanned event and it is already costing us some resources. Then an inquisitive worker wanders into the work envelope of other robots down the line while the line is stopped. There is no danger seen. The robot will either be repaired in place or removed and replaced. In either case there will be great pressure on repair personnel while the line is down. This is conducive to many types of errors.[9]

This particular robot is placed back in operation but without being synchronized with the rest of the line. No one is in the immediate area of the repaired device, and it is started up. Far down the line, hidden from view, is the inquisitive worker. The lines start up and two unplanned things happen. A robot arm, holding a large die and close to the worker, unexpectedly swings around into the head of the worker. That particular robot is not the only one acting suddenly, without warning. Two other robots, also acting strangely and out of sequence, do great damage to themselves, support equipment, and finished product.

In another case, a robot is stacking boxes and loses a large bolt out of the control device for its pneumatic hand. The bolt falls on a pallet, causing a stacked box to jiggle to one side and fall, ruining the contents. One other bolt is holding the hand in place. It comes loose later in the day and the pneumatic apparatus controlling the robot's hand swings free, flying over the barriers and striking an observer. The falling of the first bolt was not noticed by either the robot itself or the robot's software/human monitor.

Robot Mishap Causes

A small, electrically powered robot in a small work envelope and with a working load of a few ounces will not likely endanger a human much. However, a robot standing 10 feet tall, using hydraulic and pneumatic power, and with a working load of 300 pounds presents the following possibilities:

1 Its mounting may break and allow it to topple and crush someone or cause other damage.

2 Operated without fluid at proper operating temperature, the entire robot assembly may shudder and shake, which can cause parts to loosen or break. Certainly, its accuracy will be off.
3 More complexity means more chance of hoses and parts breaking or rupturing, with an immediate shift in the robot's work envelope.
4 Even operating properly the robot can crush a worker.
5 Its load is such that if it were dropped it could crush someone or something.

Some of the hazards that cause robot mishaps originate with:

1 **Configuration.** Sharp edges or corners on either the robot or product.
2 **Power Source.** Combination of power including electricity (up to 440 V), hydraulic fluid (1500 psi), or air (80 psi) under pressure. Each is a well-known danger source.
3 **Control.** Many are computer-controlled and depend on programmed instructions and/or signals from internal or external sources to trigger varied actions. With unplanned inputs they can unexpectedly change paths.
4 **Robot Error.** Robots are often highly articulated, complex devices. If any part of their systems fails, wild uncontrolled motions can result. The sources of failure are many:
   (*a*) Electrical noise.
   (*b*) Power surges or pressure drops.
   (*c*) Broken hydraulic or pneumatic lines.
   (*d*) Frayed electrical cables.
   (*e*) Short circuits.
   (*f*) Loose connections.
   (*g*) Failed electronic components.
   (*h*) Encoder and other sensor errors.
   (*i*) Dirty servo valves.
   (*j*) Excessive heat.
   (*k*) Overly humid weather.

In one factory, robot start malfunctions occur at the rate of several dozen annually. Some typical examples of robot misbehavior are:

1 In a training program the slewing shaft suddenly swung away from its preprogrammed direction.
2 The arm of a robot suddenly shot up as the oil-pressure source was cut off after the robot's work ended.
3 A robot made a motion that was not part of its program.

4 A robot started moving when its power source was switched on, although the interlock conditions were not ready.
5 When operating alone, the robot destroyed the work it was to weld because of an error in a program instruction.
6 During hot summer weather, the robot arm suddenly shot up although it had been operating normally.

The most common causes of robot mishaps are:

1 "Noise" interference.
2 Oil pressure valves.
3 Servo valves.
4 Encoder-related trouble.
5 Computer board malfunctions.
6 Errors traced to misjudgment.
7 Errors in operation by human.

One study showed a control system fault happening during every 100 hours of operation. Robot body faults occurred between every 100–250 hours of operation, with runaway robots happening between 500 and 1000 hours of operation. Undoubtedly this operational record is improving as time goes on and the systems are improved. It has been claimed that simple robots require only one or two days of maintenance per year in normal operation, but this figure does not agree with other statistics.

Training Role

One advantage of robots is that less space is needed for their operation than for humans doing the same work. Their compactness often means that there is little space for a technician to do repair work. He may have to work in an awkward posture, often lifting heavy robot parts—both are mishap sources. The maintenance person is also more exposed to hazards than others and needs special training. If the maintenance person is in a mishap the training records should be checked to see if the following were covered (this is also valid for operator training):

1 Safety training.
2 Robot operations and systems overview.
3 Programming (teaching).
4 Identification of robot malfunctions.
5 Robot justification and motivation to buy it.
6 The hydraulic system.
7 The electrical system.
8 The mechanical system.
9 Interfacing to related equipment.

10 Alignment and adjustments.
11 Troubleshooting and diagnostics.
12 Preventive maintenance.
13 Installation procedure.

The training takes several hundred hours and is good only if the worker has the proper background.

COMPUTERS

The computer is both a blessing and curse in that its ability to carry out computations far surpasses human capabilities. It also makes them beyond human verification. A plus or minus sign placed in the wrong place by the programmer can ruin the best calculations as it did in causing several nuclear plants to be rebuilt as an earthquake precautionary measure.[10,11]

What is so often overlooked is that a central goal of design should be to prevent failure. This is critical to identify exactly how a structure or material may fail. The computer cannot do this by itself. The use of artificial intelligence to make use of past experience from a file of failures is not yet practical, but extensive data on some types of mishaps are being developed toward the day this becomes practical. As yet, the engineer still must ask the critical question of whether a part will fail under certain circumstances, but the answer depends on human judgment. The computer can only act as a file clerk and answer if it has the information.

The fact that a computer makes the calculations quickly makes us tend to use it for designs in which every part is of minimum weight and strength, thereby giving the most economical structure. This was not practical with hand calculations and designers used to settle for oversize and overstrength in design to be sure of being within the acceptable range. Making every part as light and as highly stressed as the codes allow leaves little room for error in the computer's calculations, the parts manufacturer's products, or the construction or execution of the design. Thus, optimized computer designs may be the least safe designs. The growing use of micro- and minicomputers and the vast number of programs for them are a ripe source of mishaps. In innovative designs the designer may not know what questions to ask the computer. In the hands of inexperienced designers, the right questions are often not asked.

The investigator may have trouble visualizing the interaction of mishaps and problems in computer software. The software itself is not hazardous, but once associated with a system, it can be as potentially hazardous as the system itself.[12]

For a hazard to occur there must be an unwanted release of energy, energy normally contained by some hardware component(s). A failure or error in hardware control must cause or allow the hazard to occur. Thus, software that monitors and exercises control over the condition or state of the hardware is considered critical to the system from an investigatory viewpoint. However, other systems that provide indirect control or data for critical processes must also be considered, since bad data can lead to potentially hazardous decisions by human operators or companion systems.

SOFTWARE

A computer program in itself is not unsafe. Only the systems which it controls can do damage. This requires that we look at the overall system of programming, manuals, and computer hardware compatibility instead of merely looking at the computer program. Software problems fall into four broad categories: (1) an unexpected and/or unwanted event occurs; (2) a known and planned event occurs, but not when desired; (3) a planned event does not occur; and (4) the magnitude or direction of the event is wrong.

Traditionally, computer failure has been viewed only in terms of electronic component failures in reliability and safety analysis. The usual measure of mean time between failures (MTBF) cannot be applied directly to software, since software failures consist of errors in the program. These errors have three classifications:

1 Incorrect or incomplete specifications and requirements that lead to incorrect or incomplete designs.
2 Software errors made during programming or coding.
3 Hardware-induced corruption of the program.

It is not easy to detect software errors. In theory the proper testing program would detect all errors. In practice this is nearly impossible because of all the possible combinations of sequences and timing. Besides, time-compression techniques make the error window so small that it may be undetected. In the past, software was tested or debugged by trial and error. When an error was found a change was made. This system has never been good enough.

Software safety is a relatively new field requiring a team effort of all involved parties. The total system viewpoint must always be considered. Proper safety requirements must be written into proposal contracts for software.

Software errors show up in hardware-induced failures. Sometimes the reverse happens: A hardware failure can cause an error in a computer word that induces a software failure because the software instruction no longer has its original meaning. While the failure mechanism is within the hardware, the software is erroneous because it has a new and unintended meaning. Such failures may be temporary or permanent.

Hardware sensor failures can provide erroneous inputs. In an automated system, this leads to erroneous judgments, resulting in improper commands. Again the failure is entirely within the hardware, but the software outputs are inappropriate and defeat the system operation.

COMPLEX SYSTEMS

This chapter has looked mostly at independent systems, but mishaps invariably involve two or more systems. This calls for a closer look at systems.[13]

The systems view considers events in terms of relationships and integration and the interaction between several components of the system. It perceives these relationships as being dynamic and on-going. Systems thinking is process thinking, that is, how is the system as a whole working?

There are many types of natural and man-made systems.

1. **Open and Closed Systems.** When a system can exist only in a certain environment it is an open system. A closed system can exist in more than one environment.
2. **Adaptive and Nonadaptive Systems.** An adaptive system can change to fit its environment, while a nonadaptive system cannot change.
3. **Loosely Coupled Systems.** Loosely coupled systems can handle shocks, failures, and pressures for change without failing. In loosely coupled systems there is a better chance that quick, spur-of-the-moment buffers, redundancies, and substitutions can be found, even if they have not been planned for. In a loosely coupled system the operator or maintenance person has more chance to properly react, to make corrections.
4. **Tightly Coupled Systems.** These systems respond more quickly to changes, but the response may be disastrous. Tightly coupled systems allow only one way to reach a goal. Buffers must be designed in advance. In a tightly coupled system, corrections may be made, but they must be precise.

Linear Relationships, Interactions, and Limitations

Linear interactions are found in expected and familiar production or maintenance sequences. They predominate in all systems, but even the most linear process can have complex interactions.

Machines work with linear chains of cause and effect. When a linear chain fails, it is possible but unlikely that a single cause for the sequence breakdown can be found. Linear systems have minimum feedback loops and thus less chance to trip up designers and operators. Controls are near the task and are attached to special-purpose equipment. The input to run linear systems is likely to be sent directly and reflect actual operations.

The catch is that straight, linear, cause-and-effect relationships seldom exist in today's complex workplaces. Knowing the nonlinear nature of a system is critical in mishap investigation. When one or more systems fail it is because of several factors, which may have amplified each other through interdependent feedback loops.

The nonlinear nature of humans also makes a single mishap cause unlikely. Any person functions between upper and lower limits and there is a constant variation between the limits, even without any disturbance. Such a state, called homeostasis, offers the system many options to interact with its environment.

Complex Interactions

Complex interactions involve components outside the normal sequence. They relate to unfamiliar, unplanned, or unexpected sequences. They may be unseen or hard to grasp. Complex interactions with tight coupling may produce a system breakdown. We have designs so complex that we cannot anticipate all the possible interactions. Adding safety devices may only cover hidden paths in the systems.

Unanticipated interactions cannot be found by usual methods. An activity between two independent systems that happen to be adjacent can cause an interaction that is not foreseen or linear. There are also branching paths, feedback loops, and undefined jumps from one linear sequence to another unrelated sequence. These adjacent events impinge on systems that in turn pick up other new and unwanted linear relationships. They multiply, sometimes logarithmically, as subsystems are breached. These unexpected interactions may not be related to the system(s) being investigated and thus have no apparent interest to the investigator.

Interrelated systems investigation will likely be complex. The many hidden values in complex interactions discourage us from making valid findings on their causal relationships. Faced with environmental, social, and interrelationship problems in systems, some argue that these are outside their specialties, outside their control. In an investigation, we often tend to distance ourselves from these forces rather than seek relationships.

SUMMARY

This discussion of the systems approach to mishap investigation does not attempt to make the investigator an expert in investigating any special system, but instead shows how the systems approach works. The approach can be used in almost any size or type of investigation. It is not confined to large or complex investigations. It should be clear, however, that with the complexity of modern systems the investigator will seldom be able to thoroughly investigate a system or systems. If the investigator is not a specialist in the area, experts can be found in nearly any specialty.

Proper use of the systems approach calls for careful coordination with those who are working on other or connecting systems involved in the same mishap. The investigator should not presuppose the importance of any system, but should cover them all. Complete and accurate documentation through notebooks and pictures is needed to investigate the system and its components as they are removed from the wreckage.

Note that isolation of the system and components requires the investigator to note what is different about the operation of that system after the mishap. He then investigates what caused the difference or change. This process is documented several ways in this book, and an entire chapter (Chapter 10, Change Analysis) is devoted to that approach.

QUESTIONS FOR REVIEW AND DISCUSSION

1 Explain how a traditionalist might investigate an auto mishap differently from someone using the systems approach.

2 How many systems can you list in the human body? In your car?

3 How many systems can you name in a large ocean-going oil tanker? List them.

4 Bring in, for inspection, a clear example of an impact failure.

5 Inspect a clear light bulb that has failed and write a description of what you see. Bring it to class for inspection.

6 List at least eight types of robot failures and give an example of the possible results of each failure.

7 Describe five specific ways a computer could be used to aid in mishap investigation.

8 What does the term "nonlinear nature of humans" mean?

QUESTIONS FOR FURTHER STUDY

1 Secure the public information copy of any recent National Transportation Safety Board mishap report. Explain how that report documents the use of the systems approach to investigation.

2 Locate and bring in, for inspection, evidence of rotational damage from a mishap. The rotational damage must be clearly visible.

3 Secure a light bulb or the remains of one from a wrecked auto headlight. Analyze it for hot or cold damage. Bring it to class.

4 Establish contact with an organization or person equipped to investigate or furnish guidance for investigating a robot mishap. Report on your contact.

5 Research and write a concise (100 words or so) report on how software data connected with controls, navigation, or production can be retrieved after a mishap.

6 Describe at least three possible complex interactions within the meaning of this chapter.

REFERENCES

1 Systems Field Investigations Guidelines; File No 51-7494-4, U.S. Army Board for Aviation Accident Research, Fort Rucker, AL.

2 "Instructional Notes on System Failures" (unpublished), University of Southern California Faculty, Los Angeles, 1972–1979.

3 Thumann, Albert, *Electrical Design and Energy Conservation,* Fairmont Press, Atlanta, GA, 1978.

4 *Accident Investigation Procedures and Technology,* Transportation Safety Institute, Oklahoma City, OK.

5 Swallom, Donald, and Michael Cassity, "Lamp Failure Analysis—An Investigative Technique," graduate paper at the University of Southern California, December 1985, 17 pages.

6 Boroson, Harold, et al., "Guide for Protection of Electrical Equipment from Carbon Fibers," ARCOM/AMC/AFLC/AFSC Joint Technical Coordinating Group (unpublished draft), May 1978.

7 "NASA Gives Nod to Composites," *Aviation Week & Space Technology,* Dec. 17, 1979, Vol. 74.

8 Rimson, Ira, "The Impact of Fly-by-Wire on Aircraft Accident Investigation," International Society of Air Safety Investigators, *Proceedings of the Eighth International Seminar, Caracas, Venezuela,* Oct. 3–6, 1977, pp. 3–6.

9 *Working Safely with Industrial Robots,* Peter M. Strubhar, Ed., Robotics International of SME, Dearborn, MI, 1986.

10 Brown, J. Crozier, "Anatomies of Computer Disasters," *Proceedings of the First International Conference on Computers in Civil Engineering,* American Society of Safety Engineers, New York, 1981, p. 250.

11 Petroski, Henry, *To Engineer Is Human,* St. Martin's Press, New York, 1985.

12 *Software System Safety Handbook* (AFISC SSH 1-1), USAF, September 1985. Also see *Hazard Prevention* magazine, January/February 1986.

13 Ferry, Ted, *A Complete Mishap Inquiry Model* (unpublished study), Los Angeles, 1985, 35 pages.

OTHER SOURCES

"Anatomies of Computer Failures," *Proceedings of the First International Conference on Computing in Civil Engineering* 1981. Note: To permit the speakers to be candid regarding the computer disasters they described, names of people, organizations, and products were changed or omitted in the published papers.

PART THREE

Analytical Techniques

Most of the emphasis in this portion of the book is on setting the direction of the mishap investigation, analyzing causal factors, and evaluating their contribution to the process of making corrective recommendations. Twenty-three techniques are presented for consideration, some in passing and a few in detail. Many are obviously related; indeed, some have been developed as refinements or complements to existing techniques.

The supervisor who investigates a mishap in his department as an added duty has far different needs and capabilities than those of the experienced investigative group charged with looking into mishaps with important political and economic consequences. This book does not recommend one technique over another. That is a matter for the reporting organization guidelines and the initiative of the individual investigator seeking better approaches to mishap investigation.

CHAPTER EIGHT

Basic Analytical Techniques

For our purposes an investigation is aimed at mishap prevention, not fault finding. That does not mean that responsibility and accountability for the mishap should not be assigned. However, we are concentrating on facts leading to the mishap so that corrective actions may be taken, even though fault or award may be the ultimate consequence.

Several mishap analysis techniques are avialable, both simple and sophisticated. The choice of technique depends on the purpose and orientation of the investigation. Often the investigator will have no choice in the matter but to follow a "company line." A few of the more common analytical techniques are discussed, not as a recommendation or to develop proficiency in a technique, but to provide information. Other concepts that are more complex or are judged to have special merit are covered in separate chapters that follow.

DEFINITIONS

In reviewing mishap analysis techniques, we find that investigative terminology varies widely. The following definitions of some investigative terms are taken from Department of Energy literature.[1]

1 **Investigation.** A detailed systematic search to uncover facts and determine the truth of the factors (who, what, where, when, why, and how) of mishaps.
2 **Analysis.** The use of methods and techniques of arranging facts to:
   (*a*) Assist in deciding what additional facts are needed.
   (*b*) Establish consistency, validity, and logic.
   (*c*) Establish sufficient and necessary causal events.
   (*d*) Guide and support inferences and judgments.
3 **Hypothesis.** A tentative assumption made to draw out and test its logical or empirical consequences, it implies insufficiency of presentable and attainable evidence and, therefore, a tentatvie explanation.

4 **Inference.** Passing from facts to probable facts whose actuality is believed to follow from the former.

5 **Judgment.** A formal utterance of an authoritative opinion; the process of forming an opinion or evaluating a propostion by stating something believed or asserted.

6 **Probable.** Likely to be so; something that can be reasonably believed on the basis of available evidence, though not proven or certain.

7 **Recommendations.** Specific methods and corrective actions believed feasible, practical, and sufficient to fulfill the judgment of needs. In general, each need is expressed in two kinds of recommendations:

(*a*) For fixing the specific problems involved in a mishap.

(*b*) For fixing systematic problems uncovered during the investigation.

8 **Codes, Standards, and Regulations (CSR's).** Authoritative specific statements of safety measures. Regulations have governmental sources and the force of law. Standards are usually voluntary. Codes may be either regulations or standards.

9 **Guidelines.** Less specific than CSR's.

10 **Risk.** The probability during a period of activity that a hazard (see Appendix A) will result in a mishap with definable consequences.

(*a*) Original risk Arising from an unanalyzed, uncontrolled activity.

(*b*) Residual risk Remaining after an analysis and some control action.

(*c*) Calculated risk Specific, analyzed, and, where possible, quantified probabilities measuring risk in a project or activity.

(*d*) Group risk Rates and projections for a class of exposure.

11 **Error.** Any significant deviation from a previously established, required, or expected standard of human performance that results in unwanted or undesirable time delay, difficulty, problem, trouble, incident, malfunction, or failure.

SEQUENCE OF EVENTS

The sequence of factors involved in a mishap was given popular emphasis through the writings of H. W. Heinrich[2] and were graphically demonstrated with dominoes. Five labeled dominoes—ancestry and social environment, fault or person, unsafe act, unsafe condition, and injury—formed the basis for a sequence of events known as the "domino theory." The domino sequence is a modest variation of the childhood game of placing dominoes in a row so that when the first is pushed they all fall sequentially. When the first domino in our line of five, ancestry and social environment, is pushed, the others will fall in sequence, resulting in the fall of the last domino, an injury. Heinrich showed that by removing an intervening domino (taking a preventive action) the ones following would be protected from the fall, and there would be no injury.

The idea that one event follows another linearly and ends in a mishap or injury makes for easy explanation of mishaps. Experienced investigators can readily point out faults with the concept. Most of the criticism centers on the idea that many events and conditions appear randomly or over a great time span and do not lend themselves to exact sequencing. Nevertheless we find some sequencing in most analytical techniques.

Modern formats have been developed, and we widely accept the concept of a chain of events leading up to a mishap. Charting procedures covered in later chapters will show sequential action as a central theme.

The domino theory is durable. First presented by Heinrich about 1929, it has been updated by several persons. In 1976 Bird[3] used a newer sequence with five dominoes identified as lack of control, basic causes, immediate causes, the incident, and people/property/injury damage. In 1978 Marcum[4] cited a seven-domino sequence. In the fifth edition of Heinrich's book in 1980,[2] Peterson made certain the tradition is carried on.

KNOWN PRECEDENTS

Based on the precept that there are "no new causes" for mishaps, the known precedent approach says that we should look at similar prior mishaps for clues to causal factors. Strictly speaking this is not an investigative technique, but the lessons learned from previous similar investigations are one of our strongest guides for solving present mishaps.

It has been said that we may have new types of mishaps but the causal factors are never new. The investigator might argue that it is unlikely that we would have previous experience to draw on in a mishap involving testing, design, the operator, management, and so on, but each there are known precedents for each of these casual factors.

In more sophisticated investigations it is common to draw on data banks for historical data on similar events and to seek clues to causal factors related to the new mishap. Sometimes the investigator takes a computer readout of previous mishaps to the site, thus gaining direction and guidance through historical precedent. Experienced investigators intuitively draw on their prior experience in seeking solutions. They would be foolish if they did not.

ALL CAUSE/MULTIPLE CAUSE

Our historical beginnings in mishap investigation show that we have long sought the "one cause" of a mishap. This single-minded determination remains evident in many investigation report forms and procedures, and in some analytical techniques. If, in the next few minutes, a catastrophic mishap occurred, the man in the street would want to know the cause. Radio and television commentators would seek to comment on the mishap in terms of one thing that made the event possible, such as operator error, mechanical failure, or the like. This concept has been slowly replaced by the search for several causal factors, with one of them eventually identified as primary. The primary cause is the one most responsible for the mishap or the one event after which the mishap was inevitable.

The need to list all causal factors contributing to a mishap has long been known. This multiple-listing process all too often seeks a scapegoat as the main causal factor, agina focusing on one cause. A later development lists all causal factors without endorsing any particular one. This approach says that all factors leading to a mishap should be corrected to keep similar mishaps from happening; therefore none should either be given precedence or excluded. Of course, some causal factors, because of their seeming importance or because they can be easily corrected, tend to take priority positions. The "all cause" approach emphasizes the need to seek all causal factors and not to concentrate on one area to the exclusion of others. This approach is estimated to have raised the average number of correctible causal factors found in mishap investigation from four to eight: a doubling of investigative efficiency.

CODES, STANDARDS, AND REGULATIONS

Referred to simply as CSR's, codes, standards, and regulations refer here to governmental requirements related to mishap investigation. CSR's do not specify an investigation technique, but they do sometimes specify when a mishap is to be investigated and the type of data to be furnished. In this role the CSR's, by inference, provide direction for the investigation. The data needed are usually minimal, to keep reporting from being a burden; they therefore give little assurance of a thorough investigation.

When the requirements of the law have been well covered, as with the National Safety Council format for mishap reporting and recording (which also meets OSHA requirements), the data sent to the government agency are still minimal. CSR's provide, at best, information that tells us where our problems lie and, to a certain extent, the nature of these problems. They do little to help us prevent more mishaps. The unfortunate result is that most workplace mishaps are investigated just enough to meet minimum reporting requirements. Even with traffic, marine, and some aircraft mishaps the results will not provide good prevention data if there is little public interest, if the mishap is routine, or if the investigators are busy. OSHA requirements are also included in the National Safety Council procedures.[5]

Many businesses and industries belonging to the National Safety Council have developed their own mishap reporting forms incorporating or using the Council forms. Thus, the CSR requirements direct the investigation. Detailed information may be found in the National Safety Council literature.[5]

THE FIVE M'S

This mishap prevention technique is also a good investigative approach. It apparently started with the Venn diagram popularized by a beer logo, in which three overlapping circles show the relationships between man, machine, and environment. Then the term *media* replaced environment to give us three M's. The addition of management has provided a fourth M. The point is that by systematically considering the man, the machine, the media (environment), and management, most of the considerations in investigation are covered. Various admirers of this approach have made checklists of items to be considered under the four M headings. It would seem that the depth of consideration of the items under the four M's is the key to success of this approach. A fifth M of "mission" has been added to now give us five M's.

HAZARD ANALYSIS DOCUMENTATION

Modern safety practice calls for several types of hazard analyses in a process. Ranging from early design hazard analysis to on-the-job safety analysis, they are documented and kept available in files or data banks. Once the mishap investigation preliminaries are under way, reference to the documentation may aid the investigator in several ways. The documentation:

1. Indicates the proper way to carry out a task or the particular hazards identified in the task.
2. Gives the investigator a base for correct task performance so that departures from the norm may be noted.
3. Points out design deficiencies that should have been corrected or provided for in some way.
4. Indicates whether special precautions, identified by analysis, were carried out.
5. Often refers to specific managerial and supervisory responsibilities that should have been completed with the task or product, thus alerting the investigator to errors of omission or commission by that group.
6. Outlines maintenance procedures, which are frequently a subject for hazard analysis and should be checked against the task performance in a mishap.

There are many other types of supportive analyses and several ways they can be used by the investigator for direction and assistance. These few examples should give the investigator many ideas.

INFERENTIAL CONCLUSIONS

The expert investigator is often preoccupied with gathering factual information to support the identification of mishap causal factors. While this emphasis is proper, the pride in dealing "only with the facts" often leads to failure to consider information that cannot be fully documented. This passage from facts to probable facts based on the former can be a valuable investigative aid.

Bruggink, formerly of the National Transportation Safety Board, said that we must not only be able to gather facts but must also be able to draw inferences and deal with assumptions to help place our facts in perspective.[6] It is proper to gather both negative and positive evidence and continually assume that something happened to disturb the routine operation. This constant gathering and weighing of factual and nonfactual evidence is an inductive process that continually draws provisional conclusions, each of which is then objectively appraised and tested for validity. As the investigator proceeds, the inferences and evolving facts become intertwined and difficult to separate. The danger for either the new or experienced investigator is in taking the inferences to be fact after a while. Continual care is needed to ensure that inferences do not, by constant reference, become seen as fact. It is well to combine facts, theories, speculation, and conjectures in sequence to form hypotheses for evaluation. They must then be either proved or refuted on the basis of the best available information. An orderly handling of theories through the prove/refute stage is necessary.

REENACTMENT

Mishap reenactment is common and not so much a valid investigative technique as an intuitive reaction. In its most common form the operator involved in a mishap directed to "Tell me exactly what happened." In a slightly more advanced version the operator is placed at his equipment and told, "Show me exactly what happened." The most sophisticated and certainly safest version of reenactment calls for duplication of the operation in a computer simulation and/or modeling of the process under laboratory conditions.

Reenactment has some limitations. The operator may not realize what he did wrong and may not be able to duplicate it. If live equipment is used it must be emphasized that unsafe acts must not be repeated. This calls for careful, step-by-step supervision of the reenactment process. The stories of what has happened during reenactment range from ruining duplicate pieces of costly equipment to repeating the injury when the reenactment got out of hand. A person should not be asked to reenact a mishap unless he is mentally and physically up to the task. In a traumatic event the operator may never be able to carry out a reenactment, but someone may be able to take his place.

THE HARTFORD

Organizations frequently involved in mishap investigation often build their own style and approach to meet their special needs. Insurance companies who have the largest corps of investigators, have developed many excellent techniques. The example presented here comes the Hartford Insurance Company and has been selected because of its simplicity, validity, and wide use.

The Hartford procedure is based on looking at three things usually present in any mishap—equipment, material, and people—and is thus known as the EMP approach. Four verbs are listed under each of the three headings, and combined they cover nearly all mishap situations:

Equipment	Material	People
Select	Select	Select
Arrange	Place	Place
Use	Handle	Train
Maintain	Process	Lead

The conditions leading to any mishap will always involve the 12 items listed here under equipment, material, and people. To investigate all facts about a mishap the investigator must ask the why, what. where, when, who, and how of each item under the three main headings. For example, the first item is the selection of equipment. Asking the why, what, when, where, who, and how of that item and the three underneathe gives us 24 individual combinations uder equipment. The EMP model, then, provides us with a look at all the factors that might be present in a mishap as well as an order by procedure to follow. Still remaining is the need to recommend appropriate corrective action.

SUMMARY

A few analytical techniques have been presented and briefly discussed. They seldom stand alone; our investigative approach most often combines several approaches that happen to suit us best. There are many other analytical and investigative techniques. Some are given a more thorough review in the following chapters. Appendix C is a general summary of the techniques with brief comments.

QUESTIONS FOR REVIEW AND DISCUSSION

1. Review the mishap reporting form used by your employer or organization and comment on its adequacy.
2. Compare the definition of investigation and recommendations in the chapter. Are they in harmony with each other? Indicate how they are or are not in harmony.
3. Comment on definition 7(*b*) in this chapter. What does that definition mean in your words?
4. Does the four M analytical technique seem to be based on a sequence of events?
5. Does the subject of "inferential conclusions" as discussed in this chapter agree with the definition of inference early in the chapter? Does it agree with "hypothesis"?

QUESTIONS FOR FURTHER STUDY

1. How would you improve the mishap reporting form for your employer? If you insert new requirements, what is the "ripple" effect?
2. In your line of work, which type of risk—original, residual, or claculated—is most common? Give two examples.
3. Referring to the definition of risk in this chapter, how do you go about quantifying risk?
4. Explain how the known precedent technique could be used in investigating a mishap involving an exotic new metal and laser beams working together in a brand new operation.
5. Does your organization keep job safety analyses on file? If so, bring one for inspection.

REFERENCES

1. Johnson, William, *Accident/Incident Investigation Manual,* Department of Energy, Government Printing Office, Washington, DC, 1976, pp. 1-7 to 1-17.
2. Heinrich, H. W., Dan Petersen, and Nestor Roos, *Industrial Accident Prevention,* 5th ed., McGraw-Hill, New York, 1980, p. 22.

3 Bird, Frank E., Jr., and George L. Germain, *Practical Loss Control Leadership,* Institute Publishing, Loganville, GA, 1985.

4 Marcum, Everett, *Modern Safety Management Practice,* Worldwide Safety Institute, Morgantown, WV, 1978, p. 29.

5 *Accident Prevention Manual for Industrial Operations,* 8th ed., National Safety Council, Chicago, 1981, pp. 162–163.

6 Ferry, Ted S., *Elements of Accident Investigation,* Charles C. Thomas, Springfield, IL, 1978, p. 15.

CHAPTER NINE

Systems Safety/Fault Tree

An earlier chapter discussed looking at systems as a mishap investigation approach. The systems approach in that context and systems safety, of which fault tree analysis is a technique, are very different. A true systems safety approach can be complex and time-consuming. Originally developed for sophisticated aerospace systems,[1] it has the ability to clarify complex processes through charts and models. There are several systems safety analysis methods, and a complete analysis requires especially trained engineers and advanced mathematics. The system as we use it involves the personnel, equipment, and environment of a system combined to carry out a function. If they do not interact properly, a mishap results. Instead of dealing only with a mechanical, electrical, or hydraulic system, we look at its function and the relationships that exist among personnel, equipment, and environment as that function is carried out. The systems approach realizes that a failure in personnel, equipment, or environment affects the other two. We investigate the system for causal factors and remedies. Developed initially to prevent mishaps, the approach can also be used to investigate them.

GENERAL

The analysis of systems used several techniques. The ones considered here are failure mode and effect analysis (FMEA), technique for human error rate prediction (THERP), and fault tree analysis. Other methods such as program evaluation review technique (PERT) and the critical path method (CPM) are also mentioned[2] as appropriate techniques.

Failure Mode and Effect Analysis (FMEA)

The FMEA approach can be useful for investigating situations where complex, interrelated components are involved. In FMEA a failure or malfunction is seen as part of the system, and the effects of the failure are traced through the system while the effect on task performance is evaluated. The procedure assumes a complete knowledge of the system. It has the disadvantage of considering only one failure at a time, so that the possibility of other failures and their interactions may be overlooked.

An investigative application in a recent failure involved a commonly used aircraft. It was found[3] that certain propeller-reversing linkage was disengaged at the time of the mishap. Because of the worldwide use of the aircraft in transport operations a failure mode and effect analysis was conducted by the manufacturer, and it was shown how the linkage could fail or become disengaged. Under some flight conditions the propeller could then go into reverse pitch with possible catastrophic results. The result was an airworthiness directive for all aircraft using that engine and propeller.

Technique for Human Error Rate Prediction (THERP)

Although it was designed as a preventive tool, THERP can be used for investigation. By isolating each human subtask performance, human error probability can be studied. Since THERP establishes basic human error rates (BER), we can combine probabilities to consider the likelihood of human error in a particular mishap.

FAULT TREE

Fault tree analysis is credited to scientists at Bell Laboratories and was refined by the Boeing Company to analyze problems with the Minuteman missile. The technique was adopted by the Department of Defense, and it is commonly used by their contractors as a preventive and sometimes failure-analysis device. Our coverage of the fault tree is not rigorous or comprehensive, but it will demonstrate its use as an investigative tool.

A fault tree is a symbolic logic diagram that shows the cause-and-effect relationship between an undesirable event and one or more contributing causes. While discussed here mostly in relation to hardware, it can also be used to analyze software.

Of the analytical techniques mentioned for system safety, the fault tree technique is used most widely.[4] In this method an undesired event (a failure) is selected and all the possible things that can contribute to the event are diagramed as a tree. This use of a tree allows us to see where our problems lie and to place our thoughts on paper to produce a logical sequence of causal events.

Beginning with an undesired event (mishap or failure) the fault tree analysis reasons backward, tracking events that could have led us to the unwanted happening. A model (schematic) of the system is used to trace the contributory events. Symbols portray the events in a treelike network developed with increasingly significant and detailed events. In the end completely independent events are reached.

Computers, with the right software, make it possible to process large amounts of information. The computer processes information logically and in a definite sequence. If numerical values of the probability of occurrence can be developed for each of the fault tree basic events, the technique lends itself to computer programming. This is particularly interesting since the key elements of the fault tree, AND and OR gates, correspond to logic gates in computer use. This allows the probabilities of an event to be calculated. The analysis requires a knowledge of Boolean algebra and access to a computer. The calculations will not be used here, although the investigator may see the need for them and would work with a system safety analyst for best results.

Symbols

Fault tree language is largely symbolic. A few of the symbols are shown in Figure 9.1. A more complete listing is found in Appendix D.

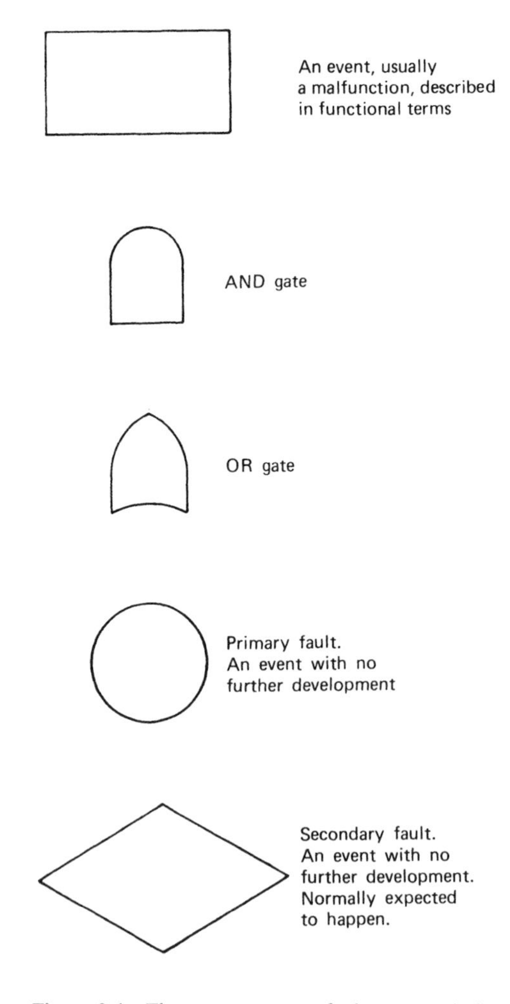

Figure 9.1 The most common fault tree symbols.

Sequence of Events

We recognize that most mishaps have identifiable events sequences linked with them. Figure 9.2 shows a typical sequence of events. It is obvious that the figure lacks details on some of the events that could have taken place. Fault tree analysis is a method of analyzing the ways and means in which an undesired mishap event occurs. Through logic network diagrams, the relationships and sequences of events that could cause the event are analyzed. The components are said to either operate successfully or fail. Partial or degraded operation is not considered; anything less than full operation is a failure.

Fault tree events have a hierarchy and can be classified as follows:

1 **Head Event.** Sometimes called the "top" event, the event at the head of a tree that is anlyzed by constructing the rest of the tree. It could be, for example, a headlight failure.
2 **Primary Event.** A primary fault in which the component itself malfunctions. Caused by a characteristic of the component itself, such as a light-bulb failure due to excessive voltage.
3 **Secondary Event.** A secondary fault or effect caused by another component, device, or outside condition such as a battery or generator failure.
4 **Basic Event.** An event (fault or normal) that occurs at the element level and refers to the smallest subdivision of the analysis of the system. This event would be at the base of a tree branch.

Our analysis, then, will go as follows:

1 Select the head event (the undesired event that is to be investigated).
2 Determine primary and secondary events that could cause the head event.
3 Determine the relationship between the causal events and the head events in terms of AND or OR.

The Top Event

If, during a mishap investigation, we wish to focus on one undesired event that took place and to indicate the complex relationships that brought about

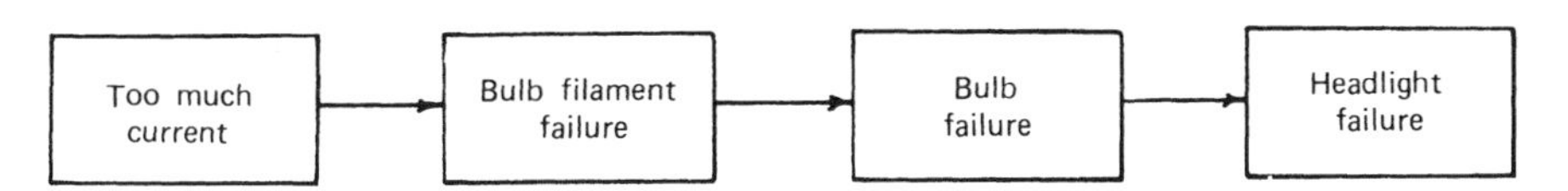

Figure 9.2 Typical mishap sequence of events.

the event, we will find that the fault tree is excellent for just this purpose. While commonly used for quantitative purposes, the fault tree can be used for its qualitative aspects because of the systematic way in which the various factors in a mishap can be presented.

The mishap, damage or injury is the top event[5] and is the rectangle at the top of the tree. Contributing factors are added and the tree is developed. Events with zero probability of occurrence are eliminated. If the top event is the failure of a piece of equipment to operate, the fault tree can then be used for troubleshooting. The tree will also show the importance of a component to operation of the product. Examples of top events are the following:

1 Damage to a machine.
2 Injury to a foreman.
3 Overheating of a chemical.
4 Collision with a lamppost.
5 Loss of control of a motorcycle.

Placing a mishap such as one of these in the top event position starts the tree development.

A related use of the fault tree for investigation is with successful operations. A successful top event is investigated to determine if a change to a hazardous situation took place. When there is no change the event is eliminated as a causative factor. Finding a change alerts the investigator to a possible causal factor.

Building the Tree

Before going ahead with the fault tree assumptions, certain stipulations must be made about the characteristics of the components, conditions, actions, and events:

1 No operation is partially successful. There are only two modes: success and failure.
2 Basic failures are independent of each other.
3 If we are going to quantify our tree, each item has a constant failure rate. These rates conform to an exponential distribution.

After the top event is selected, the causal contributory events are drawn, branching downwards. They are connected to the top event through AND and OR gates.[5] If all the events listed below it are needed to make the top event possible, the AND gate is used. If only one event is needed to make the top event possible, then the OR gate is used.

To rephrase that a bit, if an OR gate is used, the presence of any of the contributory events (probably primary events here) will cause the top event to take place. If all the contributory events must be present for the event to take place an AND gate is used.

An electric bulb is often used to illustrate a fault tree. Assume that a light bulb went out, a machine operator could not see what he was doing, and a mishap resulted. It becomes important to know why the light bulb failed at this critical time and place, so we decide to develop a fault tree on the event.

The top event in our tree is the light bulb failure. If we had a wiring diagram we could imagine that the bulb either burned out or the power failed, so we draw a fault tree and connect these events with lines. Since the failure could be *either* bulb or power failure, we connect the events through an OR gate, as shown in Figure 9.3. If the bulb burned out, it may not be to our best interest to analyze the event deeper since we are not in the light bulb business. Besides, for a few cents more we get a long-life bulb. If the bulb was all right and burned properly when power was turned on, we might want to look into the power failure by considering more events under the power failure branch of the tree.

Each contributory event is now subjected to the same process and is studied to see when and how it will occur as the tree branches downward and possible causes are shown.

Several fault tree analysts have used the case of the disintegrating grinding wheel to illustrate the fault tree process. In a slight variation, The Case of the Driller's Eye is used to illustrate the process. In this situation (Figure 9.4), a worker was using a drill press when the drill broke, and he was hit in the eye by a flying piece.

Finishing the Tree

The downward development process continues until all available information is used. The bottom level of each branch should be a failure or some other initiating (basic) event. These events are generally noted by circles or daimonds. Most analysts use the circle to indicate complete development of that branch of the tree or termination for external reasons. The diamond shows that the event cannot be pursued further or that there is a lack of information. An example of the diamond might be the bulb burning out in Figure 9.3. If this event was critical to the situation, an expert might be brought in for deeper investigation.

Limitations

Proper use of a fault tree requires extensive knowledge of the design, construction, and use of a product. The field investigator may not have time to build this expertise. He may then ignore the process or seek the services of a specialist in fault tree analysis and in the system being examined.

A fault tree is a logic diagram that shows cause-and-effect relationships. Documentation beyond the fault tree is certain to be needed. Mishap

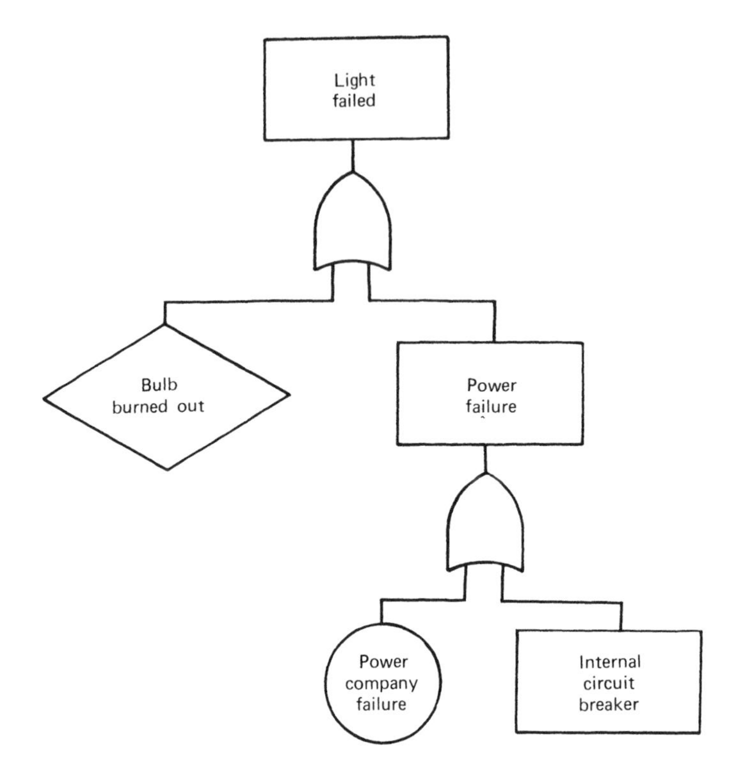

Figure 9.3 Simplified fault tree.

prevention specialists might point out that after completing a fault tree there may be good cause-and-effect information but the mishap prevention task has not yet started.

There are shortcomings in the fault tree logic, which requires either success or failure of an event. For example, it is possible for a system component to operate at reduced capacity. Or, for a human, there are many performance levels that may not be entirely successful under many conditions. Fault tree analysis does not allow such fine distinctions. Because of this and other characteristics of the fault tree process, many feel that there are too many limitations on analyzing the human role. On the other hand, many systems safety analysts hold an opposing view and feel that human performance can be analyzed within a fault tree model.

Man, Machine, and Environment Roles

It was intended that only techniques accepted and proved by common use would be included in this book, but two other approaches merit mention. this is perhaps an answer to the criticism regarding the inability of fault tree analysis to handle the human factors.

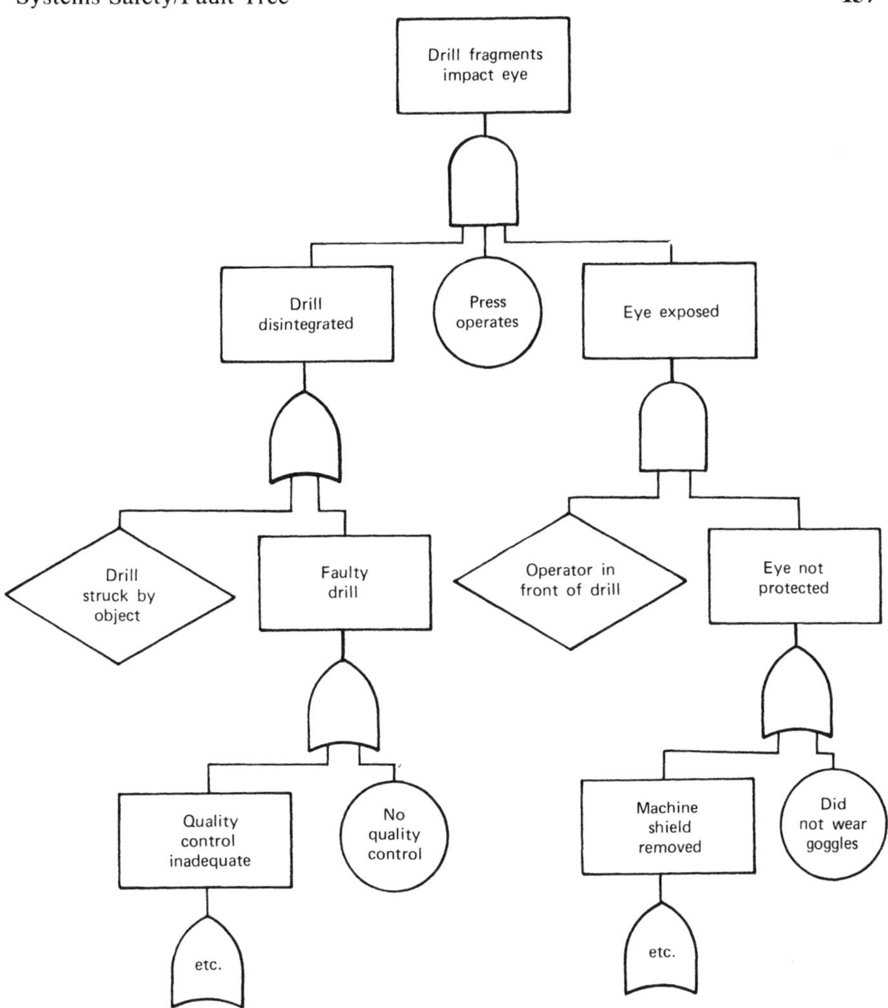

Figure 9.4 The case of the driller's eye.

Hammer[6] suggested using a human factors fault tree for investigation. He said that the fault tree can be used to separate the personnel (human), equipment (machine), and environmental factors involved in a mishap, as shown in Figure 9.5. This approach segregates events involved in a mishap, preparing them for in-depth, individaul fault tree analysis.

Nertney and Horman[7] discussed introducing the human element into the analytical methods to include fault tree analysis, failure mode and effect analysis, and other event trees. In relation to fault tree analysis, they developed a parallel with certain hardware failures. They found that because the human is subject to many more errors and more varied types of errors, the human in a fault tree tended to make the "tail wag the dog" by showing

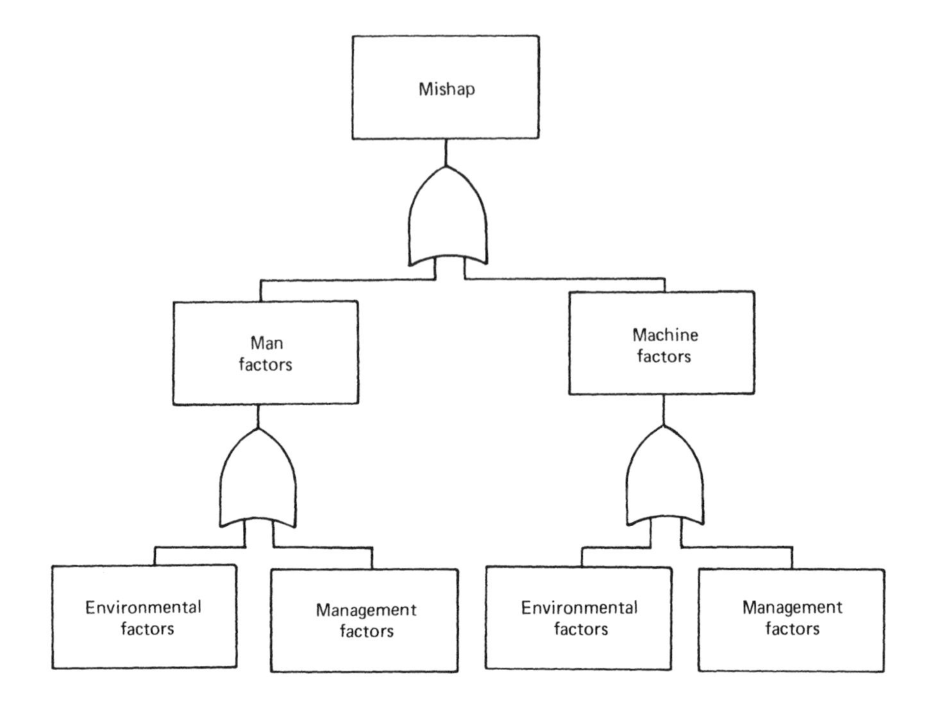

Figure 9.5 Fault used to segregate man, machine and environment.

the system far more dependent on operator errors than on maintenance or installation errors. They did not say that the human should be eliminated from the fault tree, but that the analyst should be aware of the effects of humans on the system. A similar situation was found to exist in relation to failure mode and effect analysis. They did, however, develop a human factors event tree, which is described in reference 7.

Cost Effectiveness

Cost effectiveness analysis is a method for finding how much resources we should invest in mishap investigation. It allows us to quantitatively measure the effects of various system failures. The probability of a small loss would not justify the expense of a detailed investigation. Cost effectiveness helps us to decide whether to proceed with an investigation and how far we should go with it. In the end it can help us determine the value of our investigations by measuring the ivestigation costs against probable savings.

SUMMARY

Fault tree analysis is an excellent means to graphically order the causes and effects of a mishap and provide direction for gathering information. The information that can be shown on a fault tree is limited under the best of circumstances. More information will always be needed to augment the diagram(s). The fault tree is only a diagram showing the interrelationships of mishap factors. For the expert analyst, it provides an opportunity for computer analysis, since events can be expressed in terms of Boolean algebra. For general investigation use it identifies cause factors. It does not suggest corrective action, but it helps to identify problem areas that require such action.

Fault tree analysis, like most verification processes, can be costly. The amount of effort needed to construct a large fault tree is probably too great to allow a complete analysis even with the aid of software tools.

QUESTIONS FOR REVIEW AND DISCUSSION

1 Name five techniques associated with systems.

2 Does the fault tree reason forward or backward?

3 Describe Boolean algebra and explain how it works in lay language.

4 Describe in your own words how fault tree analysis might be used in an investigation. Give an example.

5 If a brake system fails due to lack of fluid, is the lack of fluid a basic, primary, or secondary event in a fault tree analysis?

6 Draw a fault tree involving any mishap, carrying it to three levels of events.

7 Do you agree with the limitations on fault tree analysis for mishap investigation under the heading "Limitations"?

QUESTIONS FOR FURTHER STUDY

1 Bring an example of PERT or CPM for class inspection.

2 Can THERP be quantified? Describe a procedure for its quantification.

3 Little is said about cost effectiveness as an analytical tool. From other readings, describe some of the procedures used in working this out. Name the procedures. A one-line description of each is enough.

4 Draw a fault tree of any mishap and carry it to circle or diamond symbol events.

5 Find an actual example of or reference to fault tree analysis being used for mishap investigation.

REFERENCES

1 Hammer, Willie, *Handbook of System and Product Safety,* Prentice-Hall, Englewood Cliffs, NJ, 1972.

2 Grimaldi, John V., and Rollin H. Simonds, *Safety Management. 4th ed.,* Irwin, Homewood, IL, 1984, pp. 286–289.

3 King, James B., letter to Administrator, Federal Aviation Administration, "Safety Recommendations A-79-91," National Transportation Safety Board, Nov. 28, 1979.

4 *Accident Prevention Manual for Industrial Operations—Administration and Programs, 8th ed.,* National Safety Council, Chicago, 1981.

5 Hammer, Willie, *Product Safety Management and Engineering,* Prentice-Hall, Englewood Cliffs, NJ, 1980.

6 Hammer, Willie, meeting in Los Angeles, Jan. 14, 1980.

7 Nertney, R. J., and R. L. Horman, *The Impact of the Human on System Safety Analysis. SSDC-32,* System Safety Development Center, Idaho Falls, ID, September 1985.

CHAPTER TEN

Change Analysis

We intuitively seek an answer to what is different about a mishap situation from all the other times the operation has been carried out without a mishap. It may be a subconscious reaction, but we know that something has changed to make a mishap possible. We must look for that change, and it is always present. The process has been formalized for management and investigative purposes. Since it is based on change, the process is called "change-based analysis" or simply "change analysis."

BACKGROUND

The role of change in mishaps is carried back to before World War II by Johnson,[1] who later played a major role in developing an analytical mishap investigation tool based on change. Historically, change in design or revisions to operating procedures have been a focal point for those seeking to prevent mishaps. Even before World War II it was realized that a change in form, fit, or function of a part signals the need for a review of the components and subsystems (including interface) until no change in operation is detected. In a well-run organization all changes are documented so that there is a base from which to work. All sources of change are canvassed for changes so that their effect on the system can be investigated and corrections can be made if necessary.

The concept of change-based analysis was refined for the Air Force by the Rand Corporation. Two of the developers on the Rand project were Kepner and Tregoe, who subsequently wrote a book on the subject[2] and developed change-based training courses for business problems. Members of the National Safety Council staff attended Kepner-Tregoe courses and learned to apply change analysis to mishap problems. Subsequently, several companies picked up change analysis at their investigation base. Among them was Aerojet Nuclear Corporation, which, with Johnson, brought change analysis into the Management Oversight and Risk Tree (MORT) system. This led to its common use throughout what is now the Department of Energy and its contractors.

Change Signals Trouble

The alertness to change is often a case of survival in our world. In business the chief executive bases the course of his corporation on perceived changes in operation that forecast problems. Highway patrol officers are constantly on the lookout for changes as they carry out their patrol: changes that will alert them to deviant behavior, that is, a departure from the norm that forecasts trouble. The homemaker enters her home and stops, sensing some change that may spell trouble. The truck driver notices a slight change in his steering just before he loses control. A pilot feels a little roughness in the engine, a change telling him something is wrong. After a plant mishap, the investigator gets the operator to tell about it in his own words, expecting the operator to recognize a change in routine, departures from the norm that may have made the mishap inevitable.

Although change is often used here in the singular, in reality many changes, often minor and subtle, may be involved in any given mishap. Changes have a tendency to multiply, to lead to one another, and to quickly get out of hand once the first deviation or two have occurred.

Directional and Exponential

The impact of change can be better appreciated if it is seen that change is both directional and exponential,[1] as shown in Figure 10.1. Directional means that change keeps on going. Once headed in a new direction it does not change direction unless there is another change. If you expect things to return to normal after a change, forget it. This is an added burden for the investigator in his final report, when the change must be identified and corrective action recommended if a return to normal is desired. By exponential we mean that the changes interact to compound the effects on mishaps. When a driver changes his method of driving to one that is less safe, he will probably keep it up if he gets away with it (directional). Furthermore, others may join him for some apparent benefit (exponential). If a truck driver gets away with 70 mph he may never return (directional) to the speed limit if conditions permit the change. If we improve (change) a section of the highway and permit double trailers (change), the better roads and heavier loads may permit larger (change) trucks to go faster (change) with more dangerous (change) cargo. The result of the greater speed, heavier load, and highway improvement is a revised direction of activity (directional) with a vastly increased number of changes (exponential).

Many changes brought about by new technology tend to be in this cumulative, exponential exposure leading to mishaps. During an investigation the investigator must consider all possible changes and their likely exponential relationships.

The fact that a change results in a new direction and increases exponentially does not mean it is obvious. The changes that cause mishaps tend to be

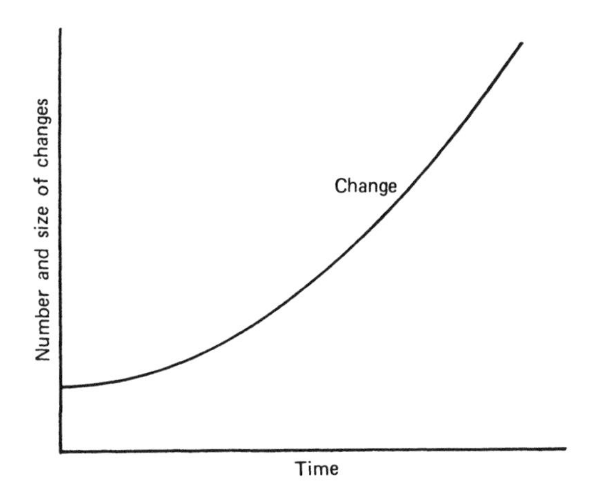

Figure 10.1 Effect of change.

subtle, obscure, slow-acting, hard to detect, and so gradual as not to alarm or alert anyone. In one typical mishap, a change made five years earlier combined with a recent change (new oil additive) to produce a mishap with undesired results.

THE CHANGE ANALYSIS PROCESS

The basic change analysis process as developed by Kepner and Tregoe is classically illustrated by Figure 10.2. Six steps are involved in the process:

1. Look at the mishap situation.
2. Consider a similar, but mishap-free, situation.
3. Compare the two situations.
4. Set down all differences between the situations.
5. Analyze the differences for effect on producing the mishap.
6. Integrate the differences into mishap causal factors.

The change analysis process is refined and visualized by use of a matrix[1] shown in Table 10.1. The matrix can be used as a worksheet to permit written entries in the various blocks. Experienced analysts take a desk-sized sheet of wrapping paper and draw in the matrix with crayon or magic marker. The 25 potential factors listed on the left side are then examined. These factors may not completely define a situation, and the analyst should not hesitate to add others. Examples of some of the changes to be made can be seen in Kuhlman,[3] where substitutions have been made.

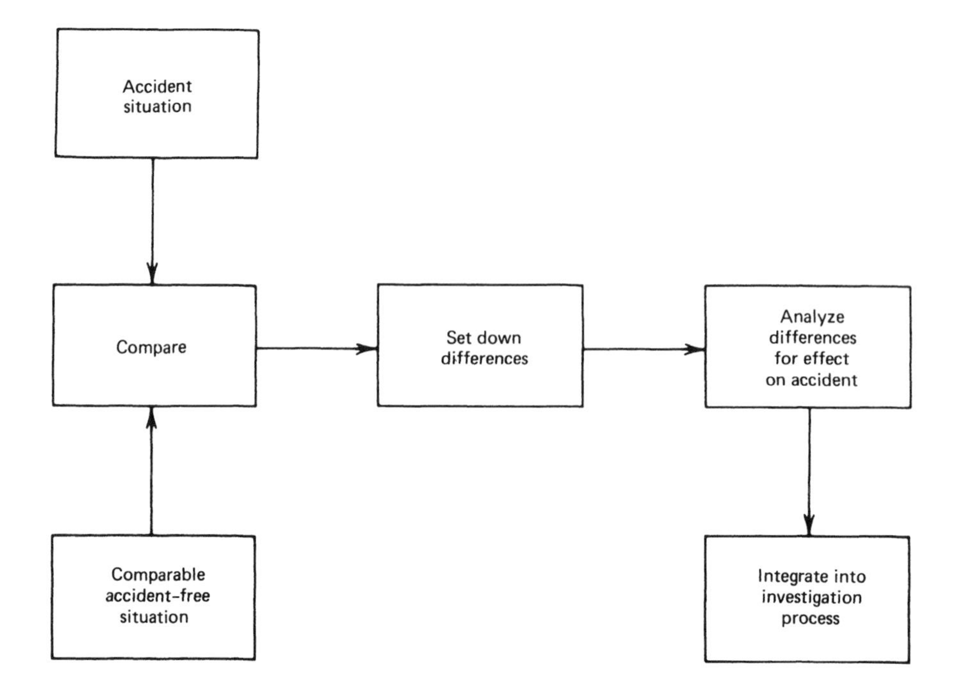

Figure 10.2 The six steps of change analysis.

CHANGE CLASSIFICATION

Changes take so many forms that the investigator may have trouble organizing his thoughts. The Management Oversight and Risk Tree (MORT) process[1] provides us with two paths to identifying change. The items in Table 10.2 were originally conceived as counterchanges but are also serve as sources of change. Whenever an item under the seven headings takes place, so does change. A review of the situation with these items in mind will reveal possible changes and also enable the investigator to quickly size up the mishap situation.

The MORT process also indicates changes by source as follows:

1 **Planned versus Unplanned Changes.** Planned changes are those brought about by affirmative actions of some type while unplanned changes are those detected by monitoring.
2 **Actual versus Potential or Possible Changes.** Actual changes are those found by reports and observation, while potential or possible changes call for analysis to see if they actually occurred.
3 **Time Changes.** This refers to deterioration over a period of time, possibly gradual, and to interaction with other changes.

Table 10.1 Change-Based Mishap Analysis Worksheet

Factors	Present Situation?	Prior Comparable?	Differences?	Affective Changes?
What				
Objects				
Energy				
Defects				
Protective devices				
Where				
On the object				
In the process				
Place				
When				
In time				
In the process				
Who				
Operator				
Fellow workers				
Supervisor				
Others				
Task				
Goal				
Procedure				
Quality				
Working Conditions				
Environment				
Overtime				
Schedule				
Delays				
Trigger Event				
Managerial Controls				
Control chain				
Hazard analysis				
Monitoring				
Risk review				

4 **Technological Changes.** These are considered whenever new projects and processes have been started, particularly those near technological boundaries that might bring new or unsuspected changes.

5 **Personal Changes.** These consider the many variables that can affect human performance.

Table 10.2 Types of Changes That Can Cause Mishaps

Substitute	Rearrange	Reduce	Modify
Power	Sequence	Omit	Color
Ingredients	Pace	Shorten	Shape
Process	Components	Split	Sound
Approach	Schedule	Condense	Odor
	Pattern		Motion
Combine		Reverse	Meaning
Blend	Adopt	Order	Light
Units	Outright	Direction	
Assortment	Related		
Ensembles			

6 **Sociological Changes.** These are related to the five types listed above, but consider the role of society in influencing the individual.
7 **Organizational Changes.** These are shifts in responsibilities or ill-defined responsibilities that leave interface gaps where "no one is watching the store."
8 **Operational Changes.** These are procedural changes that take place without safety reviews.

WHEN TO USE CHANGE ANALYSIS

The need for safety-related, change-based analysis is related to the simple fact that any "real-life" operational system is constantly experiencing changes in personnel, procedural systems, and equipment. Unfortunately, when such changes are made, the effect on the rest of the system is often not evaluated. These oversights and omissions can, and many times will, lead to mishaps.[4]

Change-based analysis techniques can aid investigation in the following areas:

- **Troubleshooting.** Often the relevant facts are quickly available if their need is pinpointed. A change-based format is an efficient way to search for additional information.
- **Obscure Causes.** During the first stages of an investigation, who knows what the causal factors might be? It is important that *all* changes and differences be identified whether they appear to make any difference or not. Change analysis helps to identify causal factors that are not obvious.
- **Identifying Those Overwhelmed with Change.** When change is not identified and controlled, it may soon produce uncoordinated activity.

Knowledgeable and competent personnel often fall apart under conditions of excessive and unexpected change in abnormal or emergency conditions.

- **Quick Problem Solving.** When problem-solving time is short and the need for remedial action is urgent, the change analysis technique provides a systematic approach for quick entry into problem solving with high credibiilty. For example, a mishap may cause doubt about the safety of a process or operation, and quick action is needed to get the assembly line back into production, the fleet back on the road, the store open for business, etc.
- **Avoiding Invalid Solutions.** Some managers and investigators have "canned" solutions for problems with similar characteristics. When a problem arises they apply the technique that worked the last time, only to find they are treating the symptoms instead of finding causes and solutions.

The change analysis technique is especially good when the mishap causes are *obscure* and when you need a *quick answer*. When you wonder, "How in the world could that have happened?," change analysis is a good place to start. The more remote or difficult the causes, the more likely that a rundown of the matrix in Table 10.1 will provide some clues. If a quick answer is needed, the chances are good that the systematic approach of change analysis will provide an answer faster than any other technique. A classic case involves a fleet of buses with frozen brakes. Hundreds of riders were stranded at a remote site. A call was made to a change analyst who quickly found that no one knew of anything out of the ordinary. Careful questioning of several people by the analyst, however, brought a possible solution. A check with the remote site confirmed that the problem source was an obscure change, the use of a different a hydraulic fluid additive. The fleet was on its way within half an hour of the first call to the analyst. Incidentally, he knew nothing about trucks, buses, brakes, or maintenance. He simply looked for changes.

SUMMARY

While change analysis is discussed here as a mishap analysis tool, it can be used to approach any type of problem. It has particular value as a preventive technique; that is, by being aware of and providing for change throughout an operation, management can carry out mishap prevention in an organized manner. In the prevention context, potential change provides an opportunity for corrective action before exponential growth of change can take place.

Change analysis, when applied to a complex situation, tends to be an involved process, but it still gets to the heart of the matter in a timely fashion. A great advantage is that it can be applied quickly and rationally to most situations, thereby assuring an organized approach that considers most aspects of the situation and prevents oversights. As with any technique, the investigator's proficiency increases with its use. Since it is well suited for both complex and simple situations, it becomes a good standard tool for the investigator.

QUESTION FOR REVIEW AND DISCUSSION

1 Carry out a change analysis on the mishap described below, using Figure 10.2 and Table 10.1. Some additional information may be found in the figures and tables in Chapter 3.

Situation

The regular night shift personnel at the warehouse were loading a truck backed up to the loading dock. A large convex mirror was placed high in the warehouse doorway to allow handcart personnel to see traffic approaching outside on the dock while still inside. The supervisor, who usually stood near the doorway checking the routine operation, was at the west end of the dock. A forklift with an inoperative headlight was approaching the doorway from the east end of the dock. An idle worker was riding behind the driver on the forklift and talking to him.

A helper with a heavy load on his handcart was going down a ramp with a slight incline through the warehouse doorway to the dock. He saw that the supervisor was not present and looked in the mirror, but because of the low light level and dirt on the mirror he could not see if there was approaching traffic. He looked into the dock and could not see any headlight glow from a forklift he thought was on the dock. He yelled for clearance from a man seated in a nearby truck but could not make himself heard due to a loud radio and noise from the idling truck engine. As he went down the incline onto the dock, the forklift loomed into sight at his left. His momentum carried him into the path of the forklift, where he was hit and seriously injured.

QUESTIONS FOR FURTHER STUDY

1 Compare the change analysis process described in this chapter with that proposed by Kepner and Tregoe in reference 2. Identify the changes that have been made to adapt their process to mishap investigation.

2 Using reference 4 (SSDC 21) and Figure 5 in that reference, diagram the forklift mishap described in this chapter.

REFERENCES

1 Johnson, W. G., *MORT Safety Assurance Systems,* Marcel Dekker, New York, 1980, pp. 55–73.

2 Kepner, Charles H., and Benjamin B. Tregoe, *The Rational Manager,* 2nd ed., Kepner-Tregoe, Inc., Princeton, NJ, 1976.

3 Kuhlman, R., *Professional Accident Investitgation,* Institute Press, Loganville, GA, 1977, p. 168.

4 Bullock, M. G., *Change Control and Analysis,* SSDC-21, U.S. Department of Energy, Washington, DC, March 1981, Vol. 77, p. 168.

CHAPTER ELEVEN

Management Oversight and Risk Tree (MORT)

The mishap analysis literature offers little insight into the mishap process. The Management Oversight Risk Tree technique (MORT) uses a logic tree format as a guide to seeking facts in mishap investigations. It is "an analytical procedure that provides a disciplined approach for finding the causes and contributing factors of mishaps."[1] Diagrams or charts are the heart of MORT. The logic tree illustrates a long series of interrelated questions. MORT, while similar to fault tree analysis, is more generalized and has several innovative characteristics. Once completed it provides high visibility to the mishap analysis process. It allows the investigator to review findings, present them meaningfully to others, alter the anlaysis as facts warrant, and record the investigative steps and results for later reference.

Most of the management oversight and risk tree analysis and supporting techniques have been developed by W. G. Johnson and the System Safety Development Center (SSDC) of the Department of Energy. The SSDC publications, which at this writing number about 40, represent the latest thinking, most experience, most research, and most of the written material on the subject. This chapter has been developed from that material. Readers should refer to SSDC publications for detailed study of the processes described here.

DEVELOPMENT

In the middle and late 1960s there was an obvious lack of methods literature dealing with mishap investigation techniques. It was found that only major mishaps had reports with enough detailed factual findings to support a rigorous analysis. Even then the analytical process was often unclear. As a result, an exploratory method of mishap analysis was developed that incorporated more thorough and systematic concepts, such as energy transfer, change, and syytem analysis. The use of a tree gave visibility to the process of analysis. It enabled the investigator to analyze a mishap, ask searching questions, and alter the analysis as more relevant facts or judgment might warrant. An attractive and important feature of the technique was that since the tree visibly displayed the analysis, the reviewer could review the report without laboriously plodding through a written analysis.

The term "fault," as used in fault tree analysis, was discarded (*a*) because it implied blame and (*b*) because the new method used general failures (e.g., management errors and omissions) on tracks parallel to specific event failures. This differs from the fault tree method, which usually applies to one event. The generalized MORT tree is much more complex, because it shows all the avenues that must be explored. If a fault tree analysis shows all the aspects checked and found either satisfactory or lacking, the fault tree may be as complex as MORT. As work with MORT progressed, management's role became clear and facts about management roles became as important as facts about the specific event in understanding where deficiencies lay and why the mishap occurred.

Results obtained by MORT in early use showed why the technique had great promise. Investigations by technically sophisticated personnel uncovered an average of 18 contributing factors per mishap. MORT analysis found these plus an additional 20 factors per mishap. Starting out as a method for investigation, it quickly showed promise as a preventive technique. The discussion of MORT in this chapter is confined to investigation.

THE MORT PROCESS

Within the MORT system, "mishap" means an unwanted transfer of energy (see Appendix E) that produces injury to persons, damage to property, degradation of an ongoing process, and other unwanted losses. A mishap occurs because of inadequate energy barriers and/or controls. The event follows sequences of planning errors that produce failures to adjust to changes in human or environmental factors. Energy, with its capacity to do damage, is essential to injury, damage, or process degradation. The general forms of damaging energy are kinetic, thermal, electric, chemical, acoustic, biological, and radiation.

Considering mishaps as unwanted energy transfers sensitizes the analyst to the dangers of energy buildup. Whether the energy produces injury or damage depends on the magnitude, duration, frequency of contact, and concentration of forces generated. This general approach was conceptualized by Gibson in 1961 and elaborated on by Haddon in 1966 and 1973. Handling the harmful effects of energy transfer is a basic prevention approach and involves identification of the energy source.

MORT is based on the concept that all accidental losses arise from two sources: (1) specific job oversights and omissions, and (2) the management system factors that control the job.

A third source, "assumed risks," once properly evaluated is not considered accidental, since we evaluated the risks and decided to accept their potential for mishap. Our reasons for not acting on known risks may vary from "There is no known preventive action" (such as for an earthquake) to "preventive cost is prohibitive for the value returned" (spending millions to prevent a mishap with minor damage or injury potential). It is also possible that the cure is more likely to result in a mishap than the cause we are trying to correct. Familiarity with MORT provides a permanent upgrading capability to find oversights and omissions that can result in mishaps.

MORT is increasingly used for other than investigating and preventing mishaps. It is specifically used as a tool to solve other management problems. There are literally hundreds of ways and places to use MORT. The Department of Energy has developed a series of allied MORT-based techniques and procedures to allow full exploitation of the process.

Three versions of MORT are available:

1 The original MORT charts of the fault tree type as shown on the chart inserted inside the back cover and as issued by the Systems Safety Development Center in Idaho.
2 An outline form of MORT,[2] not included, which can be used with the MORT charts.
3 Driessen's questionaire,[1,3] which may be used with the MORT charts (see Appendix F) or alone.

MORT Diagrams

The MORT diagram symbols are similar to those used in fault tree analysis, but not identical. The MORT diagram is a logic tree with a mishap as the first and top event and three main headings[1] or branches as shown in Figure 11.1. One branch deals with S factors, specific oversights and omissions associated with the mishap being investigated. Another branch lists the R factors, or assumed risks, things that are known but for some reason not controlled (for example, an earthquake). The third branch deals with the M factors, general characteristics of the management system that contribute to the mishap. Throughout the MORT system we encounter the term "LTA." We use the MORT logic tree to examine the adequacy of certain measures. If proper measures were not taken, we say they were LTA—less than adequate. The top event or box shown as losses can be the mishap, injury, damage, interruption of process, or loss of morale.

An analysis starts with the S factor branch (see Figure 11.2). It is the most rigorous branch to pursue and in turn leads to several other boxes with S factors. If you turn to the complete MORT chart inside the back cover and review the S factors (specific control factors) tree, you will see how the logical downward flow considers many more S factors. Also, scanning the rest of the chart will help you grasp the general method of operation.

The general management system (M factors) is shown in Figure 11.3 as separate from the process that produced the specific mishap background to

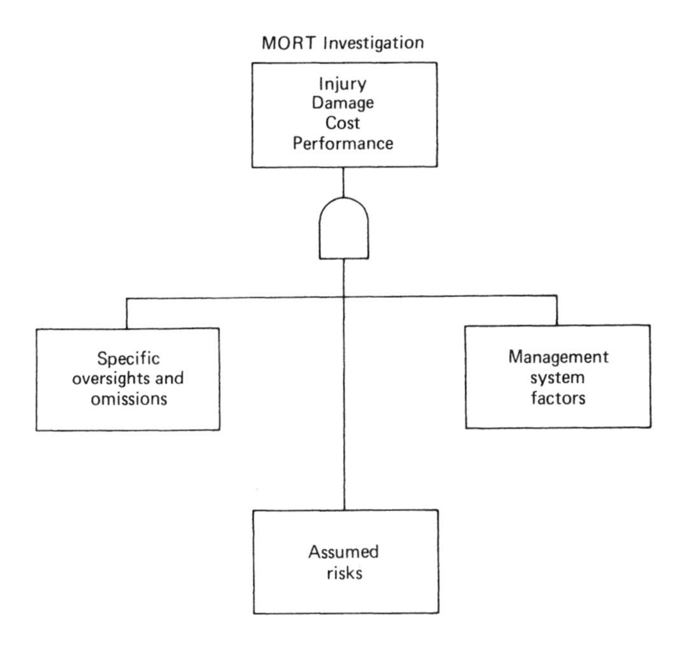

Figure 11.1 The basic MORT investigation tree.

be examined. The specific event may suggest management systems that may be less than adequate. An extended discussion on management failures is given in Chapter 14.

This chart works best if it is used as a working paper. Informality in going through the chart in the proper sequence allows for a freewheeling approach if the downward and to-the-right sequence is observed. Following the diagram will discipline the thought process. MORT serves as a screening guide and a working tool, which eventually leads to a finished report.

One point should be made clear. MORT is not usually considered a field technique in the sense that it is carried to the actual mishap site by the investigator. It is better as an analytical tool used to review and order facts brought in from the site. Shortened and improved versions of MORT make field use somewhat more practical. If the "field" is an office at or near the site, then the MORT evaluation can properly take place in the field. Although MORT has great value in ordering thought and the investigative process, its greatest value is in properly analyzing the already uncovered facts and pointing out the need for detailed scrutiny of some aspects of the mishap.

How much time does it take to carry out a MORT analysis? The developer has indicated that a review of highly complicated technical events can be carried out by a proficient analyst in about 4 hours. For less serious events,[1] as little as 20 minutes may be needed. As one repeats the analysis on different events, a sense of discipline and greater efficiency does develop. It

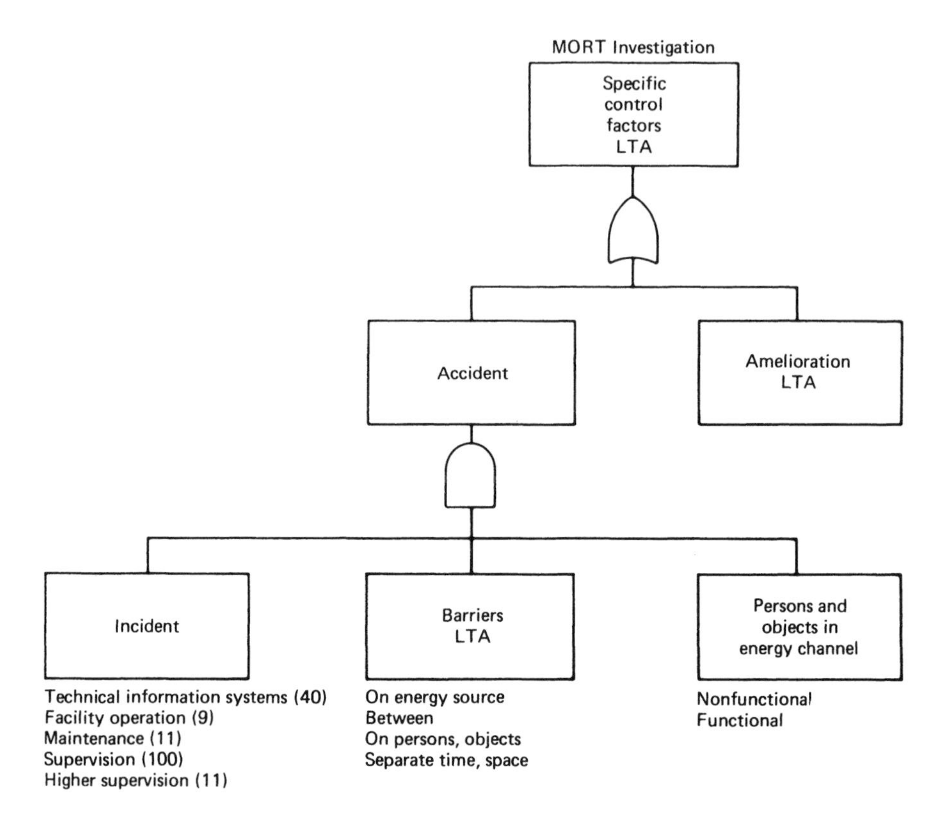

Figure 11.2 S Factors. Number in parentheses is approximate number of items to be explored in that area.

is true that a high degree of familiarity is needed to use MORT quickly and efficiently. This familiarity is developed within the Department of Energy through short seminars, while "MORT investigators" have another seminar of practice sessions in MORT evaluation and the use of other investigative tools. More commonly MORT is used as a working tool that starts with the first gathering of facts and grows as the investigation continues. Used in this manner MORT identifies uncertainties and leads to deeper investigation into those uncertain areas.

In fairness it should be pointed out that the MORT technique is sometimes taught in short order, in 1–3-day seminars. Unfortunately, people do not learn MORT well in these "all talk" seminars. It has been found that firsthand use of the tool and a high degree of familiarity are necessary to develop proficiency. However, simply reading this chapter will give the investigator many ideas on systematic approaches to investigation.

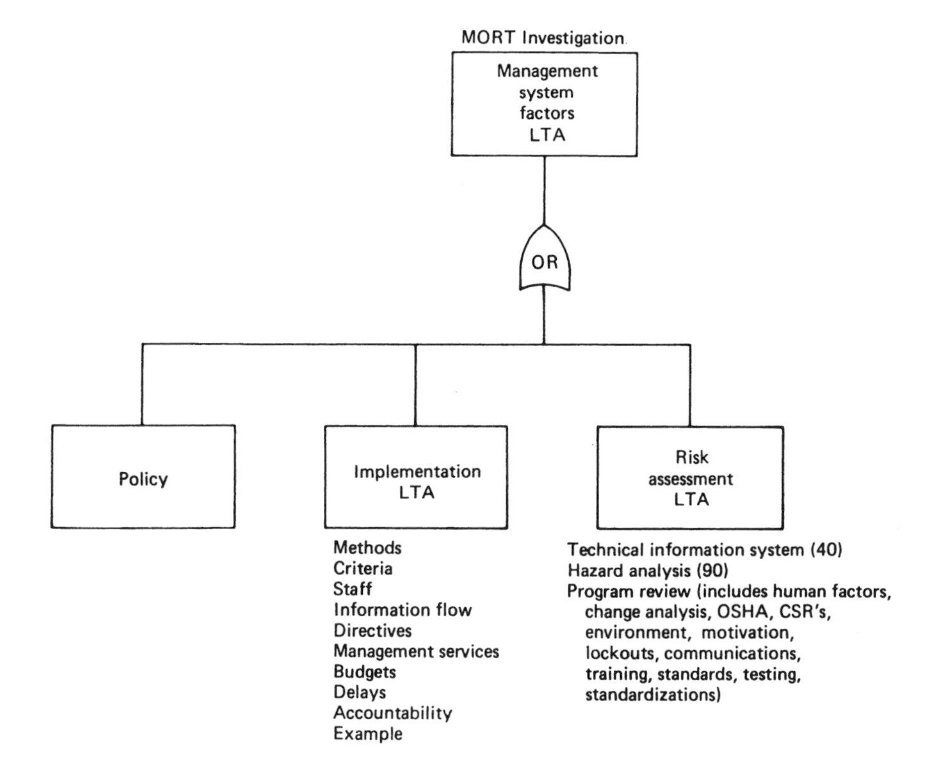

Figure 11.3 M Factors. Numbers in parentheses is approximate number of areas to be explored in that area if LTA is indicated.

Mini-MORT

A technique called Mini-MORT[4] has been developed to assist in the logical presentation of results derived from a full MORT analysis. The original MORT diagram is modified in three ways:

1. Elimination of the complete system of transfers.
2. Truncation of the long multitiered developed lines.
3. Combining factors that were analyzed individually.

The result is a simplified diagram that can be used as a basis for oral and written reports related to a full MORT analysis. The Mini-MORT may be color coded in the same way as the full MORT tree and provides a focal point for summarizing and discussing the mishap under investigation.

The Mini-MORT has not been designed to replace the original MORT diagram for analysis, since many important areas are not reviewed in enough detail for serious or major mishaps. However, the readings on the Mini-MORT chart are the same as on the regular MORT chart.

Graphics

If the symbols used in the graphic presentation of MORT seem familiar, it may be because their fault tree roots[5] are clear. As used in MORT, however, there are many differences, some of them major. A comparison of symbols is shown in Appendix D. For easy reference there are three "event" symbols: the rectangle, circle, and diamond. These are joined by two types of logic gates, the AND gate and the OR gate. As in fault tree analysis, an AND gate means all events below must happen at once for the event above to take place, while the OR gate calls for one or more of the events to take place.

The circle describes the failure of a basic component or fault. The rectangle is an event resulting from the combination of more basic events. The diamond is an event that is not developed further for some reason, such as lack of information or lack of consequences. Working down the tree then, in typical MORT fashion, a rectangle would always call for more downward development of the tree with more basic events shown as circles or diamonds. Another graphic symbol, the scroll, shows an event we may expect to occur and after evaluation we have decided to accept it as is. If risk is involved it is considered an "assumed risk." For further explanation, see Appendix D.

EVENTS AND CAUSAL FACTORS CHARTING

The events and causal factors charting sequence is an integral tool in the MORT process and has been used many times as the focal point in MORT-based investigations. The technique graphically depicts the mishap from beginning to end. This is done by showing the relationship of individual events in a mishap sequence and the related causal factors and conditions impinging on these events.

Background

Even apparently simple mishaps tend to be complex,[6] with many causal factors and with long sequences of errors and changes leading to the various events. This makes it essential to have a system, a method for breaking down the entire sequence of events into individual events with supporting information. Flow charts and diagrams have been useful for this purpose. Diagrams and charts can effectively organize the mishap data and help to confirm or deny findings. They can also aid in preparing the mishap report and may appear in the report in summary form as an aid to understanding.

Diagrams and charts have long been used in formalized investigation, but it was the NTSB that pioneered their use as analytical tools and brought them into common acceptance. Though they built extensively on the work of others, Wakeland and Benner should be given credit for originating techniques that combined the concepts in formats that are commonly used by the NTSB. Benner's special contribution is covered in Chapter 12. The use of events and causal factors charting as an investigative analytical tool has been complemented by its use as a device to help evaluate evidence.

Description

Several methods for charting events and causal factors have been developed.[3] They vary in complexity and sophistication but are typified by the simple diagram shown as Figure 11.4. Common sense is needed to develop the diagrams, but there are good reasons[7] for using a standardized format. It helps the investigator develop a scheme of operation and provides a standardized approach for those who review the report.

A diagram's existence should not lock you into that diagram as it develops. On the contrary, sequential notes will probably have to be redrafted several times during an investigation as new information appears.

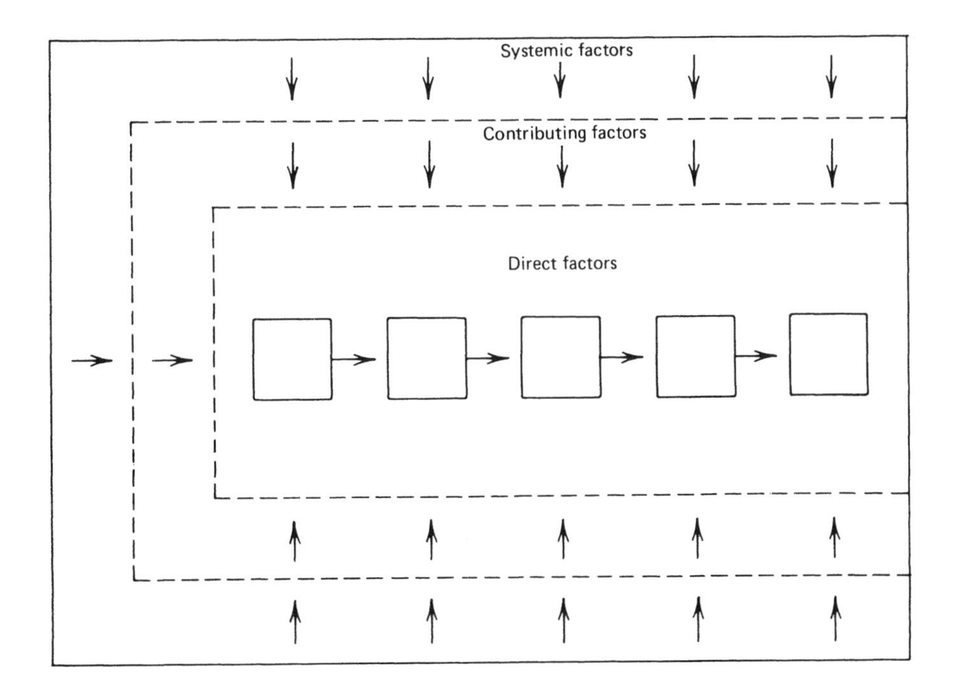

Figure 11.4 Events and causal factors charting concept.

As in MORT, the rectangles denote "events." There are definite criteria used to describe the individual events:

1 An event is an occurrence, not a condition. It is not a state, circumstance, issue, or result.
2 Each event should precisely describe a single occurrence using only one subject and verb, for example, "operator turned ignition switch to 'on' position," not just "operator started motor."
3 Events should be quantified where possible, for example, "operator increased speed to 25 mph."
4 Events range from the beginning to the end of the mishap sequence, and thus each should arise out of the previous event. This also requires defining the start and end of the mishap sequence.

To properly mesh the events (rectangles) and conditions (ovals), a format is essential (see Figure 11.5):

1 Arrange all events on a time line from left to right and connect with solid arrows. When the time of the event is not definitely known, it should be placed in sequence as accurately as possible.
2 Show the primary or main sequence of events in a straight horizontal line. Secondary event sequences are straight lines parallel to and above or below the primary sequence.
3 Each event and condition should be based on solid evidence. If this is not possible, draw the symbol with a dashed or dotted outline, and connect conditions to other conditions or events with a dashed line.

The following hypothetical case study illustrates the charting process.

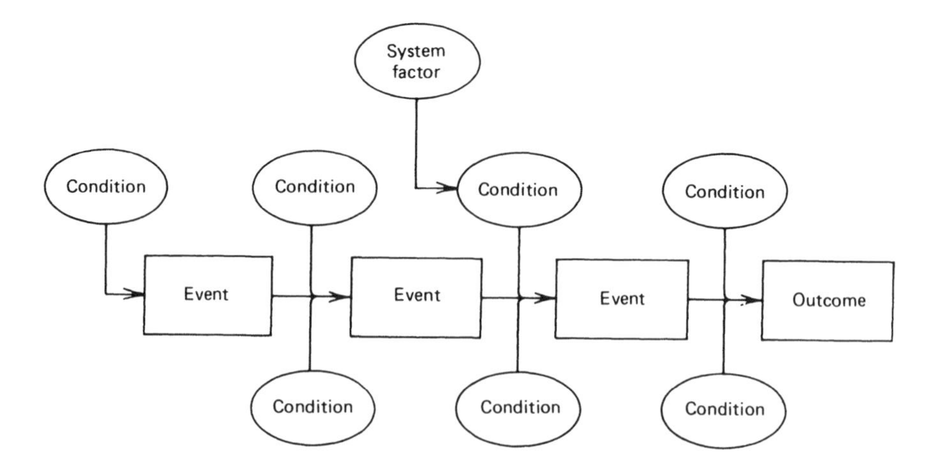

Figure 11.5 Sample arrangement of events and conditions by symbols.

Stephen was trying out a car. He picked it up from the seller, fastened his seat belt and shoulder restraint, and pulled out onto the main street heading east. The street was four lanes wide with a solid yellow double line in the center. He approached a gentle curve to the left in the far right lane. A speeding car approached from the other direction. The speeding car with four youths, Bill at the wheel, lost control as it approached the curve at about 60 mph. Bill's car went across all four lanes and struck the curb on his left, causing him to careen back into Stephen's car, now almost stopped. Bill's car struck Stephen's almost head-on. Two passengers were ejected from Bill's car; Bill and the other passenger were trapped in the wreckage. Stephen was trapped but managed to climb out a window with the aid of bystanders. All passengers and both drivers were severely injured and required hospitalization. The investigation showed that Bill's car, while 10 years old, was in proper mechanical and operating condition. Bill was not licensed to drive the car. He was not of legal age and had not received any formal driver training. The car belonged to the father of a passenger in the car and was being used without permission. Figure 11.6 illustrates this case, but in a simplified manner.

Roles and Benefits

Charting events and causal factors produces specific benefits for the investigator:

1 It serves as an aid in developing evidence, in detecting causal factors, and in determining the need for in-depth analysis.

2 Charting illustrates the role of multiple causes involved in a mishap by bringing them into the sequences.

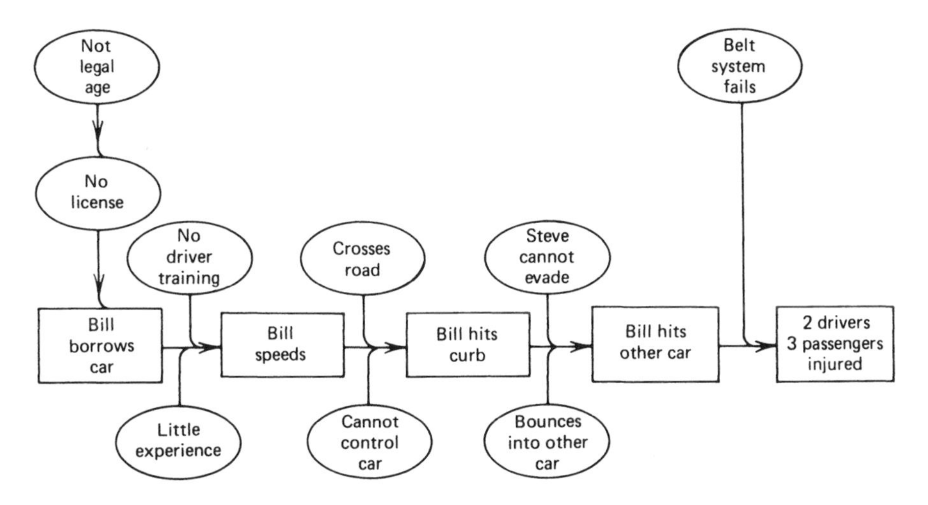

Figure 11.6 Sample application of events and causal factors charting.

3 It visualizes the sequence of events in time and the interactions and relationships of conditions and events.
4 Charting aids communication, interpretation, and summarization of the mishap.

When used with MORT diagrams, causal events charting allows verification and further analysis of deficiencies identified in the MORT process.

Causal events charting plays an important role in discovering cause-and-effect relationships without specifically placing blame, although responsibilities will certainly be allocated later. Clarifying responsibilities and reducing errors can be aided through charting by:

1 Providing a cause-oriented explanation of the mishap.
2 Making clear areas of responsibility.
3 Enhancing objectivity in conducting the investigation.
4 Providing an effective training tool for future investigation.

In addition, some contributions are made to the overall investigation goal of preventing future mishaps.

Reporting

The investigation report is meant to clearly and concisely present the results of the investigation. Charting events and causal factors aids this effort in several ways:

1 Reveals gaps in logic that lead to faulty conclusions.
2 Alerts the investigator to the possible need for more analysis when event blocks have no firm causal factors.
3 Provides items that are the basis for conclusions.
4 Provides a check between conclusions and facts uncovered.
5 Provides a basis for corrective action when it is part of the report, and allows the recommendations to be evaluated in the light of revealed events and causal factors.

SUMMARY

There are several diagraming and charting techniques used in mishap investigation. Most bear some visual relationship. Events and causal factors charting is so closely related to the MORT process that it has been included in the same chapter. MORT originates in the fault tree process (Chapter 9) and incorporates change analysis (Chapter 10). Multilinear sequencing, described in Chapter 12, is also included in some MORT processes. Thus we see MORT as a sophisticated investigative analysis tool using other proven techniques. In its entirety, MORT offers the most penetrating and through approach to mishap investigation in use today.

QUESTIONS FOR REVIEW AND DISCUSSION

1 What is a "tree," and how does it lend visibility to a process?

2 Which is more complex, the fault tree or the MORT tree?

3 What is the role of energy in mishap causation?

4 MORT is based on the concept that all accidental losses arise from two sources. What are they? Give an example of each.

5 What is meant by S, R, and M factors?

6 Is MORT a field technique? Explain.

7 LTA means "less than adequate." How would you define "adequate" for MORT purposes?

8 What methodology closely related to the MORT process is often used as a focal point in mishap investigation?

9 Explain the role of direct, contributing, and systematic factors in events and causal factors charting.

QUESTIONS FOR FURTHER STUDY

1 On the average, how many causal factors might a good investigation uncover?

2 Referring to the book by Johnson,[2] explain the concept of "assumed risk."

3 The MORT system and many others have been brought into active use through the work of the Systems Safety Development Center in Idaho. Research this center and explain how it plays a dominant national role in safety.

REFERENCES

1 Johnson, William G., "MORT: The Management Oversight and Risk Tree," *Journal of Safety Research,* March 1975, Vol. 7, No. 1, pp. 4–15.

2 Johnson, William G., *MORT Safety Assurance Systems,* Marcel Dekker, New York, 1980.

3 Johnson, William G., *Accident/Incident Investigation Manual,* GPO/NTIS, Washington, DC, July 1976, pp. 4-11, 4-23.

4 Fillmore, D. L., and John D. Cornelison, *Mini-MORT Presentation Guide,* November 1986.

5 Knox, N. W., and R. W. Eicher, *MORT User's Manual,* ERDA 76-45-4, SSDC-4, NTIS, Washington, DC, 1976, pp. 9–10.

6 Nertney, Robert, and Milton Bullock, MORT Seminar, Las Vegas, ERDA, Dec. 2–9, 1976.

7 Buys, J. R., and J. L. Clark, *Events and Causal Factors Charting, SSDC-14,* DOE/SSDC, Idaho Falls, 1978.

CHAPTER TWELVE

Multilinear Events Sequencing

Several charting concepts are covered in this book. One difficulty with most charting techniques is the lack of a time scale to show the chronological relationship of events. The multilinear events sequencing (MES) approach does order events on a time-line basis. A developer of the concept, Benner[1] has shown that the use of a checklist or extended diagram that searches for all causes assumes an extensive knowledge of the mishap process. This may discourage a search for factors not in the checklist or on the diagram. He has developed MES to a point that overcomes this disadvantage.

Pioneer work at the National Transportation Safety Board (NTSB) resulted in the regular use of sequence diagrams as analytical tools in mishap investigations, particularly those involving hazardous cargoes. The MES process brings out all factors that must have occurred for the mishap to have taken place. The MES diagram, then, aids in evaluating the evidence and helps users understand the reasoning process.

THE BEGINNING AND END OF A MISHAP

The successful mishap investigation depends on knowing precisely when the mishap began and ended. Otherwise how could we precisely determine who and what are involved? Without this precise definition we cannot develop specific corrective actions to prevent recurrence of the mishap. For MES purposes the mishap begins when a stable situation is disturbed and a mishap sequence begins. Benner[2] provides a great service with that definition, since limiting or defining the scope of the investigation depends on knowing exactly when a mishap begins. By common agreement the mishap ends with the last injurious or damaging event in the continuing mishap sequence. If the person (or thing) involved in the sequence adapts to the disturbance, the mishap is averted. Such an avoided mishap is often called a near-miss or incident. It matters little that injury or damage does not result. We believe the MES technique has great value for looking into anything that has gone wrong, whether it be a loss of morale, a delay in production, or an injury/damage situation.

The concept that a mishap begins when a stable situation is disturbed is acceptable. The extent to which that disturbance is to be investigated may be subject to question. One solution is to investigate as far back as is feasible to take corrective action. Assume, for example, that a machine operator decides to increase his output by deactivating a safety device. The investigation will seek factual reasons for this action by digging as deeply as it appears that corrective action can be taken for the discrepancies found. This may be much farther than expected. While not stated in Benner's work, an accepted approach is to look for causal factors up to 72 hours before the event. If none are apparent, the search back through time for contributing factors is dropped.

THE TIME LINE

The time line[3] is a scale that parallels the sequences of events to show a time relationship of the events that happen in a mishap. The time line may not even be shown in a final report, but during the analysis process the presence of the line helps the investigation and analysis to identify all possible causal factors. The first event that disrupted the stable situation sets the starting time of the mishap. At times this will be shown as the onset time or T_o, while the ending time is called T_n. Another way to locate T_o is to look for the beginning of the act that had to be adapted to, corrected, detected, or otherwise changed for the course of events to have a different outcome.

T_n can also be viewed as the last consecutive harmful event connected directly with the mishap, at the site where it occurred, and involving the persons at the site. If the ambulance taking a mishap victim to the hospital is wrecked on its way, this is a new event and new mishap. If a fire results from the mishap, T_n could be after the injured are taken care of and the fire has been suppressed. This may sound a little involved to show when the mishap started and ended, but knowing the exact limits of a mishap is a key to multilinear events sequencing.

THE EVENT

An events is something significant that takes place, usually stemming from one or more other actions and often leading to still other events. When a mishap occurs during an activity, that activity is composed of a series of events. Some events are sequential and some are on parallel tracks leading to the same outcome.

Something must bring about the events. We term that something (which does not have to be human) an "actor," and the act performed by the actor is called an "action." More precisely we have:

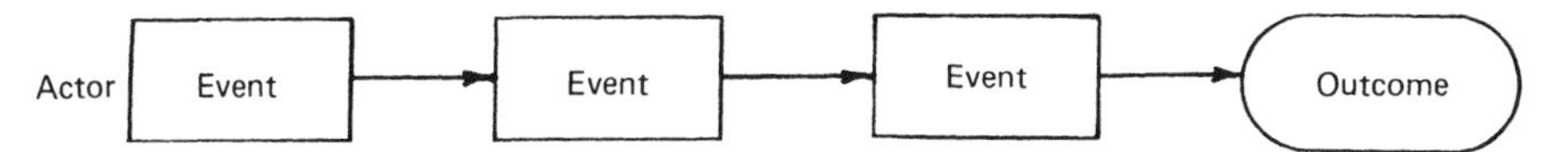

Figure 12.1 Activity events and outcomes for one actor.

$$\text{Event} = \text{actor} + \text{action}$$

To deal with specific causal factors we must deal with specific events that cause the mishap. This is done in MES by describing an event as a single action by a single actor. This calls for breaking each sequence of events into single, precisely described events, and is a main benefit of the MES technique.

SEQUENCING

If we have one actor, one outcome, and a series of events leading to this outcome, it could be illustrated as in Figure 12.1.

If we introduce another actor and the associated events, it might appear as in Figure 12.2. The spacing of the events approximates a time sequence as shown by the time line.

Conditions that influence the events can be inserted in the time flow in logical order to show the flow relationships. This is shown in Figure 12.3.

In the MES process the investigator must account for each action of every actor who (or which) brought about a change of state in the sequence. Thus each event is broken down until one event equals one actor and one action, no more. The event is reduced to its simplest form. For example, the process of applying brakes might appear to be the driver's foot pressing on the brake pedal. Actually it would be shown as several events starting with the foot on the pedal and a further series showing the actions of various parts of the braking system until deceleration ends in the vehicle stopping or impacting. Events are posted in strict sequence so that each action leads clearly and distinctly to the next, from left to right. If each event is not fully explained

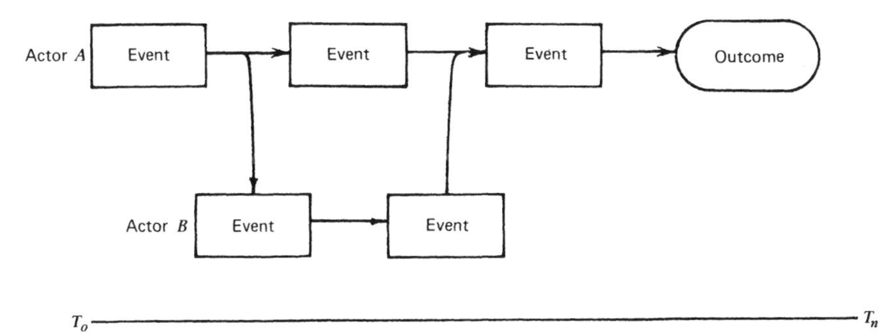

Figure 12.2 Activity events and outcomes for two actors.

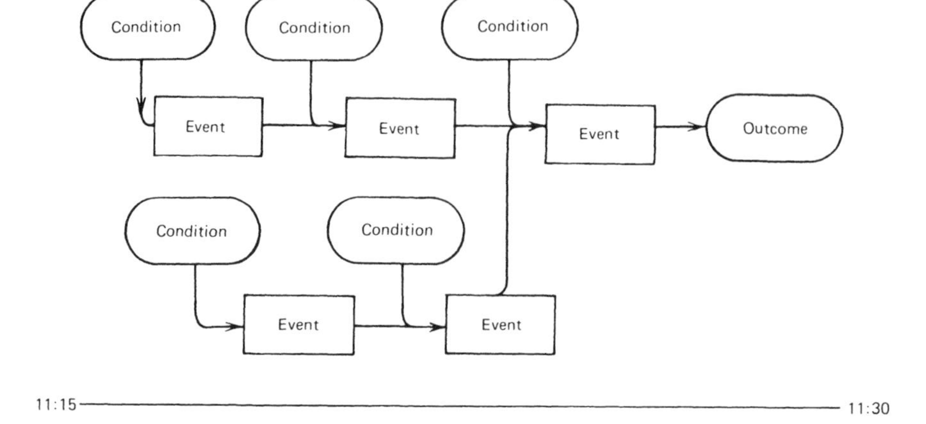

Figure 12.3 Activity events and outcomes for two actors and including conditions.

we have left out part of the sequence. If there is a gap in the understanding of an event, it is because the event to the right has not been properly broken down and explained. It shows we have not delved far enough into the sequence of events and the actors and actions involved. This total analysis may take the form of reasoning, a fault tree, or some other analytical technique. Frequently it will lead to more investigation.

THE ARROW CONVENTION[3]

The arrow convention refers to the use of arrows for linking the events and actors. Arrows always flow from the earlier event to the later, and from left to right. The arrows always flow from the earlier location of an actor to the later. An arrow could not flow from a man in the water to a bridge when he has fallen off the bridge. As this arrow convention is applied to events, actions, and conditions, gaps in the flow of events will appear, forcing discussion and review to bridge the gap. The investigator is forced into more knowledge. Arrows must not be used unless a clear sequential connection without any gap in action or event can be shown. Thus when events are displayed, the absence of an arrow shows that an explanation is needed. Events must also be tested vertically for their relationship to other actors and events.

RECORDING EVENTS

Each event is a building block that must be used correctly to make it useful for further steps and analysis. There are no shortcuts to the process.[4] The closer one can follow the steps below, the better the chances of success.

1 Changes occur often when describing events. A pencil and 3 × 5 cards are suggested for recording each event, one event to a card.

2 Always enter the name or description of each actor first. Avoid plural nouns and refer to the doer of the action by name.

3 After the actor, enter the action verb in the active voice, past tense. The action verb describes an action that started a change of state and may include "and." Only one action verb at a time is used, although it may be further described with adjectives to be specific. A separate card is required for each action.

4 In a corner of the card enter the assumed time the event began or ended so that it can be placed in the proper sequence. In a complex situation with many actors and events, time may be reduced to seconds or parts of seconds. The starting time is most important, so that the influence of the event on the sequence may be properly seen.

5 The source of the event information is often given on the card for greater precision and later recall.

6 If the events are only speculated or expected, use a different color to show that status.

7 Transfer the final card arrangement to a refined process chart.

The use of cards is necessary, because actions and events are constantly explained and shuffled to meet changes and to keep the investigation ordered and the flow logical. The final product is a mishap logic chart with the events, actors, and actions sequentially placed.

The multilinear sequencing diagram, while easy to grasp as separate events and actions, may appear complex, but that is only because all details are shown. If the events concerned with each actor are shown on a horizontal time line and there are several actors, then several rows of events may be displayed. Their interrelated arrow lines will be superimposed and connect them together. Even a seemingly simple mishap can be complex in analysis. Logic charting allows each actor, each action, and each event to be shown with their interrelationships. Thus, while the chart may seem complex, it is the apex of simplicity; each separate event is fully explained in the simplest of terms and pictures. MES is at its best as an analytical tool when the investigators have gathered enough facts to warrant at least a draft narrative report. When the events are divided into their basic elements and each subelement equals an action and actor, any need for more information or investigation will be seen along with an understanding of the mishap process.

The "acid test" of any investigation comes when the investigator asks: "Given the facts produced as events, actors, and actions, can I reproduce exactly the same outcome?" If the answer is "yes," and there are no gaps in the information, the investigator can be certain that he has done his job and can proceed with the flow chart.

Countermeasures

With a completed logic chart in hand, the investigator also has an outstanding tool with which to identify preventive countermeasures[5] that can be taken to prevent the same type of event from happening again. The first step is to examine the mishap process for ways to reduce the attendant risk.[3] The next step is to examine the events, one by one, to see where changes (countermeasures) could be introduced to alter the process. This is done by examining every linking arrow to see if the relationship between the actors can be improved in any way. Use the logic chart for this and superimpose numbered diamonds showing countermeasures on each arrow. MES gives a disciplining tool for the investigator and structures the search for options. As the investigator goes through each arrow, discovery of countermeasures is directed and developed. The countermeasures device helps managers who must take corrective action. Further discussion is beyond the scope of this book.

S-T-E-P, SEQUENTIALLY TIMED EVENTS PLOTTING

An entire volume[6] has been devoted to Sequentially Timed Events Plotting, or S-T-E-P, as it is known. The proper use of S-T-E-P procedures yields a new outlook and a revised approach to mishap investigation. Because it is also heavily based in "events and causal factors charting" and "multilinear events sequencing" covered in this and the previous chapter, it is only briefly described here.

Developed over years of experimentation and practical application in every kind of mishap, S-T-E-P is a comprehensive approach. (The terminology used in this brief overview of the field investigation application has been introduced in Chapter 11 and in earlier sections of this chapter.) When you first arrive at the scene of the accident, your first task is to document the ending states of things and people involved in the accident. This becomes an objective of your investigation—to be able to reconstruct how that state came to be.

In order to do that, you identify the actors, and photograph or sketch t_{end}. As you are documenting t_{end}, you should be separating actors from reactors and changes that occurred after the last harmful event from those that occurred during the accident process. Determining t_o is more difficult, but it is necessary so that you have a starting point for your documentation of changes that occurred during the accident.

As you document t_{end}, you develop building blocks, noting the sources of your information. These are the starting points for identifying data sources. Your primary data source will be people, whether they were participants, victims, observers, programmers, or hangers-on and volunteers. Each has a special kind of information that will help you describe the accident. Things, both stressors and stresses, can also be data sources, although they are used less often than people. When you first arrive on the scene, you should be careful not to introduce changes to the actors who are there; remember the "do no harm" rule.

After you have tentatively documented t_{end} and identified your initial data sources, start acquiring accident event data bits. If you start by interviewing a witness, try to make a mental movie of the witness's observation and actions as you listen. Sometimes you may find it useful to break events down either by breaking down an actor into two or more actors or breaking down an action into two or more actions. This involves looking for more details about what happened before, after, or between the known events. You are looking for actions that initiated other actions as well as their sequence, timing and effect.

Each bit of data that you acquire should be transformed into a building block when you can get it. Building blocks should be as complete as you can make them, since they are the basic ingredients from which your mental movie script is developed. Where information is not known, a question mark is used to indicate more data is needed. . . . [Techniques for transforming information from people to building blocks are described.]

The S-T-E-P worksheet serves as a specially structured, dynamic file for the events data acquired during an investigation. Careful attention to placement of each building block in the correct actor row and in the column that corresponds to the correct sequence of that action in relation to other actions by that actor and by other actors is an easy test of the accuracy of the worksheet.

The S-T-E-P worksheet is largely a place to lay or pin the building blocks that have been developed. It is a vehicle for organizing, storing, and analyzing building blocks. Each identified actor is listed along the left side. A t_o and t_{end} line is placed along the top of the worksheet. Building blocks are then placed in the proper column according to the time they happened and in the row across from the proper actor. Figures 12.2 and 12.3 illustrate the basic nature of the S-T-E-P worksheet.

SUMMARY

This chapter describes the multilinear events sequencing process in only enough detail to show its value and general scheme of operation. As in any logic charting process, formal instruction and frequent practice are needed to attain and maintain proficiency. One to two days of formal instruction will make you fully conversant with the MES process. However, for any investigative process, there is much value in realizing the need to break every event into subactivities and to identify the exact actions and conditions involved with each. The disciplining of the investigator's thought process through familiarity with MES is the major benefit.

The S-T-E-P procedure, to a large extent, builds on multilinear events sequencing and events and causal factors charting terminology and processes. That, however, is only a small part of a far larger and very complete mishap investigation process. A complete presentation of S-T-E-P is beyond the scope of a single chapter.

QUESTIONS FOR REVIEW AND DISCUSSION

1 Why is it so important for the investigator to know exactly when a mishap begins and ends?

2 Benner has defined a mishap (accident) as a "process." Look up two other definitions of mishap (accident) and see if his definition is compatible.

3 How far back should we go in investigating a mishap process? Why go that far?

4 Refer to the warehouse mishap described in Chapter 3 and 10. Designate a time line with T_o and T_n for this mishap.

5 Define "actor" and "action" as they are used in MES.

6 Explain the necessity to account for every "action" of every "actor" in the MES process. Explain by example the consequences of not doing this.

7 Illustrate the events you used in question **6** using the technique described in this chapter under the heading "Recording Events."

8 Does the completed MES logic chart really help to identify preventive countermeasures? Explain.

QUESTIONS FOR FURTHER STUDY

1 A worker is climbing a wooden ladder when a weakened rung breaks and he falls to the pavement, breaking a leg. Diagram the mishap using Figures 12.1 and 12.2 as guides and inserting new information as needed.

2 Illustrate the use of the "arrow convention" in your answer to question **1**.

3 What is the difference between t_{end} and t_n?

4 Using reference 6, below, turn to Appendix F on page 396, "Using MORT in STEP Investigations." Should this appendix read instead "Using STEP in MORT Investigations"? Provide a short rationale for your answer.

REFERENCES

1 Benner, Ludwig, Jr., "Accident Investigation: Multilinear Events Sequencing Methods," *Journal of Safety Research,* June 1975, Vol. 7, No. 2, pp. 67–73.

2 ———, "Accident Investigation: Multilinear Sequencing Methods," *Society of Air Safety Investigator's Forum,* Fall 1975, pp. 14, 25–27.

3 ———, *Four Accident Investigation Games,* Lufred Industries, Inc., Oakton, VA, 1979.

4 ———, "Accident Theories and Their Implications for Research," *Quarterly Journal, American Association for Automotive Medicine,* January 1979, Vol. 1, No. 1.

5 ———, "Accident Theory and Accident Investigation," *Proceedings of the Annual Seminar,* International Society of Air Safety Investigators, Ottawa, Canada, 1975, pp. 148–154.

6 Hendrick, Kingsley, and Ludwig Benner, Jr., *Investigating Accidents with S-T-E-P,* Marcel Dekker, New York, 1987, pp. 136–137.

CHAPTER THIRTEEN

Technic of Operations Review (TOR)*

TOR was conceived by Weaver[1] as a diagnostic training and mishap prevention tool but can also be used as a mishap investigatory technique. It differs from most tools in that it is directed at finding management oversights and omissions instead of operator or hardware factors. Managers tend to see the problems created by a mishap, not the underlying causes. TOR analysis centers on the system operation and its problems. Most managers are interested in this. TOR is a sequential, step-by-step process whose goal is the efficient operation of a system.

The value of TOR has been proved by Wausau Insurance Companies and their policyholders in over a decade of use. It finally became commercially available when Weaver copyrighted The New TOR Analysis. It is included here because of its potential value to the process of mishap analysis.

AN INVESTIGATIVE TOOL

In mishap investigation TOR will:

1. Review the relevant management and supervisory factors uncovered in the investigation.
2. Methodically consider, then accept or reject, more management and supervisory factors through a directed, sequential process.
3. Automatically consider contributions of supervision, management, and staff to the mishap process without pointing fingers or direct blame assessment.
4. Allow in-depth probing of difficult-to-detect management contributions.
5. Force an in-depth look at portions of the investigation not carried out in enough detail, to resolve acceptance or rejection of a TOR factor.

* This entire chapter was reviewed and updated by the TOR analysis originator, D. A. Weaver, in March 1987. TOR analysis materials are available from the FPE Group, 3687 Mt. Diablo Blvd., Lafayette, California 94549, who are the sole distributors.

The reader should examine the TOR analysis sheet (inside the back cover) at this point to see what the TOR sheet and the TOR process are all about.

THE STEPS OF TOR ANALYSIS

What triggers a TOR analysis? A mishap is investigated, and certain facts are found. The facts are then analyzed using TOR analysis and a TOR group of concerned middle managers. As with any group, a leader is needed. Experience has shown that for best interaction the leader should be selected by his own small group, rather than being named by management. This sets the stage for a free exchange of information in an atmosphere of cooperation. The TOR leader is not an enforcer, but one who can keep the analysis moving toward consensus while encouraging everyone's participation. The leader should be one who can keep the group working together and resolve difficulties and differences of opinion.

Once the TOR group leader is selected, the four steps of TOR analysis can get under way. These steps are State, Trace, Eliminate, and Seek.

State

Whoever knows the most about a mishap should briefly describe what happened. If there is a mishap report, a summary of the report and findings should be read. The group may ask questions until everyone is satisfied with the information they have. The group is now ready to start the Trace step using the TOR sheet (see inside the back cover of this book).

Trace

The New TOR analysis sheet centers on the management and supervisory factors in an operating system. The sheet displays an array of operational errors organized under eight headings. Analysis begins by selecting a single number as the prime (or main or proximate) operational error that caused or permitted the mishap to occur. Discussion may be vigorous, but the TOR leader must press for a consensus. If the mishap involved damage or injury, scrutiny may be directed to block 7, "Personal Traits," but, by consensus, the group may choose to select a starting point to any other block.

For example, the group may decide that the prime operational error was 36, "Failure to investigate and apply the lessons of similar mishaps." With 36 chosen as the prime error, the TOR sheet points to numbers 26, 43, and 61 as possible contributing factors to the mishap. As each operational error, on discussion, is judged to be a contributing factor, the TOR sheet points to additional factors to be considered. Thus, starting with one prime cause, the Trace step produces a growing list of probable causative factors, each to be discussed and accepted or rejected—did it or did it not contribute to the mishap? Discussion is driven by "yes–no" decisions arising in an orderly sequence.

The Trace step is a quick overview. By a series of brief "yes–no" discussions the group has been led through a review of its own organization and its role in the mishap. The Trace can be kept on track by using the Trace Guide shown below. Better yet, use the Trace Guide incorporated into the TOR analysis sheet (inside the back cover). Both methods yield the same result, but the latter is more convenient. Even without experience, a TOR group will quickly master the Trace guide by holding to the following brief instructions. The aim is to identify the operational errors that contributed to the *particular mishap* under review, the factors that caused or permitted it to occur. Select the prime causative factor, the START number. Circle it.

Trace Guide—The New TOR Analysis

10	20	30	40	50	60	70	80
11	21	31	41	51	61	71	81
12	22	32	42	52	62	72	82
13	23	33	43	53	63	73	83
14	24	34	44	54	64	74	84
15	25	35	45	55	65	75	85
	26	36	46	56	66		86
		37	47				87
			48				88

Proceed with the trace step, which points to other factors that may have contributed to the mishap. Judge each factor in turn, deciding whether it is in the list of contributing causes or out. If out, put a slash through the number. If in, draw a circle around it.

In brief, circle "in" numbers; cancel "out" numbers. The circled numbers will be your list of contributing causes.

To stay on track, put a small check *under* each number that the Trace process points to. Then consider each checked factor in any order. When you run out of check marks, the Trace step is finished.

The circled numbers are, in your opinion, the operational errors that caused or permitted the mishap to occur. More discussion in the Eliminate step may reduce the list to manageable proportions, culminating in the final step—Seek Feasible Corrective Action.

Eliminate

The Trace step ends with a list of six or ten, or more, operational errors, more than can be dealt with at once. This is the moment for in-depth consideration to reduce the list of factors to manageable proportions. Some may be eliminated, after consideration, as being insignificant. Others may loom large and beyond the authority of the group to cope with. Some may be self-correcting when they are merely exposed and discussed. Others may really lie in upper echelons or other departments, and appropriate communications can be discussed. Some may demand changes well within the authority of the TOR group, and discussion slips insensibly into the next step.

Seek Feasible Corrective Action

With problem areas revealed and reviewed, the idea of corrective action becomes spontaneous. The Eliminate step and the Seek step flow together into problem solving, but the tendency of people to spout solutions before they define the problem has been countered. Now the TOR group is ready for the final step, to clarify and specify the actions to be taken.

INDIVIDUAL OR GROUP?

TOR analysis can be carried out either by an individual or by a group. A person working alone, as many safety investigators must, will find the system of value. The steps should be followed as presented here for maximum benefit. The group process is favored where possible, because the group interaction that is forced on the members is a major benefit of TOR.

Group Suggestions

While the group and individual approaches are similar, the greater value of the group process deserves some attention. Here are some suggestions:

1. Small groups have better interactions. Groups of three to six persons are good. Make two smaller groups instead of a larger group if over six persons are available.
2. The group can be the investigative committee, an inquiry board, supervisors, staff reviewers, or managers who have a stake in the process.
3. The group should choose their own leader who functions as a coordinator, a friendly leader who moves them toward consensus. He should press them forward on the TOR process to a satisfactory conclusion.

Value of the Group Analysis Process

Group interaction has the advantage of a multidisciplinary and interdisciplinary approach to identifying and solving the mishap problem. The individual skills of the members can be brought to bear during the TOR analysis. Once the problem areas are identified, the critical steps of preventive and corrective actions will be easier and more readily accepted if the action is the result of a united group recommendation. The group may even include those who must decide on the preventive actions and/or carry them out.

Usually only one investigation can be conducted due to the cost and inconvenience. It should not be a surprise to find that different investigators might correctly and appropriately find different causal factors and make different recommendations based on identical information from the same mishap site at the same time. Indeed, in those military situations where two independent investigations are required it is not unusual for different findings to be produced by the two groups or two individuals. It may not be practical to make two complete investigations, but sometimes it is acceptable to have two or more completely separate analyses of the same mishap. TOR analysis make this possible. We should accept that even the most skilled investigators and reviewers can, if given exactly the same facts, come up with different conclusions and recommendations. This is not a slur on an investigator's ability. We all see things differently, and given the same facts we come to different conclusions. This is a forceful sign that there is much advantage to be gained from having different groups looking at the same facts. The conclusions of all groups may be valid and worth implementing.

One study[2] of investigation effectiveness cited the benefits of the TOR process. Over a period of 7 years, about 1500 safety professionals were presented the same facts on a mishap situation through the medium of a 27-minute film. They were then given a TOR sheet and instruction in TOR analysis. Working as groups of four to six, in classes of 25–30 people, they were asked to analyze the facts presented in the film by doing a TOR analysis. The four to six groups in each class never completely agreed. Each group usually came up with 8 or 10 causal factors. Rarely did they have as much as 25 percent agreement, even though they all had exactly the same factual information. Sometimes there is no agreement at all. Most commonly they agreed on one or two out of the 8 or 10 factors presented by each group. All their findings were usually valid. They just saw the situation differently. This tells us that given all facts uncovered through expert investigation, the critical action may be the correct assessment of causes and corrective action taken on these causes. More specifically, for the relatively small cost of a TOR analysis, we can increase our mishap investigation leverage tremendously through a multiple review of the facts and follow-up corrective action.

SUMMARY

All organizations have operating errors that arise from obscure underlying causes that TOR analysis can expose. The search for these causes is often met with fear and emotion instead of an objective search for understanding. TOR analysis changes this climate. Emotions stay cool when the group traces number sequences instead of having the finger pointed at them. The action centers on the analysis of the system, and the group gains insight into problems of the system.

TOR analysis exposes areas where corrective action is needed. It does not propose solutions, but if well done it exposes problem areas. Someone must still deal with corrective actions. It may be within the scope of the TOR group to make recommendations. They are in a good position to do so, having examined the mishap carefully. If the supervisors and managers involved in the group are stakeholders in the investigation, then it will also be their job to implement their recommendations. This is an ideal situation.

QUESTIONS FOR REVIEW AND DISCUSSION

1. What are the eight major divisions of the TOR sheet?
2. State a few rules for an individual to follow in using the TOR analysis process.
3. Divide the Seek step into at least three substeps.
4. Where would you start the TOR analysis (what number) for the mishap described in Chapter 3? Why that number?
5. If the mishap described in Chapter 3 had happened in a small organization of 150 employees, who might be part of the TOR analysis process?
6. Describe the relationship between the Trace and Eliminate steps.
7. Describe the relationship between the Eliminate and Seek steps.

QUESTIONS FOR FURTHER STUDY

1. What divisions might be added to a new TOR sheet beyond the eight now shown?
2. Do an individual TOR analysis on the mishap described in Chapter 3.
3. Review reference 1 and note the differences between a TOR analysis for prevention and one for investigative purposes.

REFERENCES

1. Weaver, D. A., "TOR Analysis: A Diagnostic Training Tool," *ASSE Journal*, June 1973, pp. 24–29.
2. Ferry, Ted S., "Techniques to Improve the Effectiveness of the Capable Investigator," *Proceedings: Annual Seminar of the International Society of Air Safety Investigators*, ISASI, Montreal, September 1979.

CHAPTER FOURTEEN

Where Did Management Fail?

INTRODUCTION

That eminent investigator, Bruggink, set the stage for this chapter when he cited several definitions of mishaps (accidents) in a widely reproduced speech.[1] Some of these definitions follow:

> Except for Acts of God, every accident, no matter how minor is a failure of the organization. (Andrews)
>
> Accidents are only managerial excuses for operational errors that result from managerial failures. (Pope)
>
> An accident is almost by definition a mark of inefficiency. (Sharp)
>
> Accidents always have a significance beyond themselves. They are always symptomatic of disorder in a particular dynamic system. (MacIver)

Many persons have recognized the need to consider the organization or a system as being at the heart of mishaps. D. A. Weaver developed the TOR concept discussed in Chapter 13. William Pope organized a society (The National Safety Management Society) composed of safety professionals who have followed that line of thought. Grimaldi[2] and Petersen[3–5] have written texts on the subject, as have others. William T. Fine, while working for the Naval Surface Weapons Center,[6] developed a technique that stems from recognizing that "all accidents and hazards are indicators of management failures; and that these management failures are causing many productive losses as well as accidents." It is Fine's particular approach that is the basis of this chapter.

HYPOTHESIS

Investigations, properly conducted, show us the management failures that contributed to a mishap. These failures may be simple inefficiencies, errors, or oversights. They often exist for years, continually causing huge losses, without anyone really being aware of them. Mishaps can be readily analyzed to detect these management failures by systematically asking one question: Where did management fail? The analytical technique presented here calls for accepting the following hypothesis[4]:

> In the investigation of any mishap, there can always be found some degree of management involvement or activity that might in some way have prevented the mishap. Therefore it is arbitrarily assumed that management will be responsible for the causes of every mishap, and the existence of every hazard.

To properly understand this process we must clarify the two key words: "management" and "failure."

"Management" refers to all levels of supervision higher than the immediate supervisor. It includes the administration and support activities, such as personnel, logistics, and maintenance. Note that the immediate supervisor is not included: The widely accepted concept of the immediate supervisor as the "key man" is not accepted here. The responsibility is considered to rest with higher management and staff.

"Failure" is any act or omission that can cause or contribute to a mishap, to the existence of a hazard, or to the commission of an unsafe act. This failure can also be called an error, an oversight, or an inefficiency.

APPLICATION

This technique is applied by investigating each mishap with a general management introspection. In each case, for every causal factor uncovered, for each failure found, and for every hazard that relates to a particular mishap, we must ask, "Where did management fail?"

The task is to determine whether any management elements took action that allowed an unsafe condition or failed to take an action that could have prevented it. Sometimes the action will be obvious. Sometimes it will always be there. There may be several such actions. The following case studies, presented to illustrate the process, are based on actual occurrences, changed slightly to facilitate the investigative intent.

Case 1: No Belt Guard

A machine operator caught his shirt sleeve in the nip point (a point on moving machinery where fingers can be caught, or nipped) between a drive wheel and a moving belt. He managed to keep his arm out of the machine by pulling it away and tearing the sleeve from his arm. Normally this event would have ended with the operator being warned about wearing long sleeves on the job and a written report showing that a guard should be fabricated for the machine. The investigator, however, decided to find out where management had failed in this minor case. He found that this was a new machine. The supervisor felt that it could not present a serious hazard, because this was the way it had been delivered. More pursuit of the management roles and answers to the question "Where did management fail?" brought this information:

1. Safety requirements had not been specified in the purchase order for the machine.
2. Purchasing did not as a rule consider safety; they simply ordered in compliance with purchase orders.
3. The safety department should have been monitoring purchase orders.
4. There was a lack of coordination between safety and procurement.
5. The operator was aware of a suggestion system, but he did not view it as a proper vehicle for reporting this hazard, which he had called to the attention of his supervisor.

The faults were quickly corrected, but some interesting side benefits emerged from this line of investigation.

1. The lack of coordination between safety and purchasing had placed an undue burden on supervisors, who were working with several unsafe machines and pieces of equipment. It was brought to light that many portable power tools did not have ground wire systems because of this oversight, and they were repaired or replaced.
2. Supervisors could now concentrate more on their tasks with proper equipment, several hazardous situations were found and corrected, and some possible mishaps were averted.
3. Better communications between various organizational sections resulted in generally more efficient operations.

Case 2: The Unsafe Scaffold

An employee walking down a hallway was bumped in the groin by a heavy plank being carried by two workers. The investigator, in tracking down the mishap causes, found that the two men were carrying the heavy plank to set up an improvised scaffold, something they did frequently. The dangers of being bumped in the groin by a passing plank are obvious, but a more dangerous situation was found connected with the scaffold. The workers carrying the plank frequently used planks, beams, ladders, and ropes to set up makeshift scaffolds. These scaffolds were generally shaky, sometimes high above floor level, difficult to erect, and time consuming to set up and disassemble. They required frequent adjustments while in place. The task was performed on the average of once a week. Asking the question "Where did management fail?," the following was uncovered:

1 The potential for serious injury existed in carrying and setting up the scaffolds but had not been noticed by management or staff personnel, even though it had been going on at least once a week for over a year.
2 The scaffold was dangerous to erect, and once erected was dangerous enough to use to offer the chance of severe injury.
3 A new easy-to-assemble scaffold had been requested, but the request had been turned down by middle management as too expensive.
4 It took three or four men about 3 hours to set up or take down the scaffold, an unproductive use of time.
5 Work frequently stopped so that the scaffold could be tightened or adjusted.
6 A new scaffold would cost $2000, while the annual savings would be around $30,000 and productivity would be greatly improved.

The obvious benefits from management investing in a new scaffold reached even deeper than that.

1 It was found that the same management bottleneck that precluded buying the new scaffold was responsible for several other shortsighted actions that prevented what would have been wise investments. A few corrective purchases greatly improved overall efficiency, and a substantial cost savings materialized.
2 The process of moving the new scaffold was far safer, and the safety and operating efficiency of the crew was greatly improved.
3 As a result of counteracting the "penny wise, pound foolish" syndrome, general plant safety and efficiency were greatly improved.
4 Morale improved visibly, and management's image was cleaned up by what was perceived to be a management interest in the worker. It was also good business.

Case 3: Little Wheels

By now it should be clear that asking the question "Where did management fail?" has a big payoff. It might be a larger payoff with more spectacular mishaps, but the results of looking after the minor mishaps can be most revealing and prevent the catastrophic ones. Consider the case of the little wheels on a two-wheeled loading cart, just like the one that might be used to take the garbage to the curb once a week at home. The cart was used to transport large loads of material around the grounds, across gravel and grass, and up over curbs. When an employee using the cart reported to first aid with a sore back, the safety supervisor realized this had happened a couple of times before with other employees using such a cart. With these back sprains costing an average of $24,000 each, it was obviously worth looking into. He found the following:

1. The operators had asked management several times for a new cart but had been turned down, because the cost of a new one, $400, was thought excessive.
2. Three men had suffered back injuries with two small two-wheeled carts in the past year, two with this one cart.
3. The small wheels did not permit carrying the loads over gravel and grass without aid, and it took three men to lift a load over the curbs several times a day. The cost of the detours and extra help came to nearly $6000 per year plus the lost time and general inefficiency.

What happened then? When the investigation showed the cost of the shortsighted approach, the following occurred:

1. Management approved the purchase of two new replacement carts for greatly improved job performance and savings.
2. A good portion of the back injury problem was eased by the advanced design of the new carts.
3. The word spread that management was expected to see these things for themselves and suggest corrective actions. They did, and there was a significant increase in morale and improved operating efficiencies as a result of improved equipment and fewer hazards.

This all happened because someone persisted in asking, "Where did management fail?" Fine documented many more cases, some of them quite dramatic[4] in the use of this investigative technique. Such failures can be found every day in nearly any organization. Management errors, oversights, and omission will be found in nearly every first aid injury and hazardous situation noted by anyone. Correcting the management deficiencies that allow these things to exist will benefit many other operations, not just the one under review. Such actions will eliminate hazards that have not yet caused a mishap but could at any time.

BASIC MISHAP CAUSES

The reader can appreciate that simply asking the question "Where did management fail?" makes the technique more basic than it really is. There must be some order to the process of finding where management failed. In this technique we start with a mishap or hazard and track it back to management errors or oversights. This does call for some expertise and judgment.

Below are 15 basic mishap causal factors traceable to management failure. One or more of these will be responsible for, or found in, nearly all hazardous situations.

1. Poor housekeeping.
2. Improper use of tools, equipment, and facilities.
3. Unsafe or defective equipment and facilities.
4. Lack of proper procedures.
5. Unsafe improvised procedures.
6. Failure to follow prescribed procedures.
7. Job not understood.
8. Lack of awareness of hazards involved.
9. Lack of proper tools, equipment, and facilities.
10. Lack of guards and safety devices.
11. Lack of protective clothing and equipment.
12. Exceeding prescribed limits, load, speed, strength, and so on.
13. Inattention, neglect of safe practices.
14. Fatigue, reduced alertness, and hypnosis.
15. Misconduct, poor attitude.

When one or more of these basic factors leading to a mishap have been identified, we are ready to link them with management failures.

MANAGEMENT FAILURES

For an example of how the causal factors listed above are tracked to management failures (responsibilities), see Table 14.1. This shows the first cause on the list: poor housekeeping. Using the example of an employee tripping and falling over equipment left in the aisle, we identify possible underlying causes of:

1. Unrecognized hazards.
2. Inadequate facilities.

We might find that the hazards were not recognized because the supervisor did not consider the items in the aisle a hazard and did not remove them. The facilities might be considered inadequate because there was not enough storage space. Looking in the third column of Table 14.1 we see that possible management failures are connected with supervisory training, supervisory safety indoctrination, and planning and layout.

Table 14.1 Causes of Management Failure

	Immediate Causes	Possible Underlying Causes	Management Failures
1	Poor Housekeeping Example: Employee trips and falls over equipment left in aisle	Hazards not recognized Facilities inadequate	Supervisory training Supervisory safety indoctrination Planning, layout

Table 14.2 Causes of Management Failure

	Immediate Causes	Possible Underlying Causes	Management Failures (Inadequacy)
1	Poor housekeeping	Hazards not recognized	Supervisory training and safety indoctrination
		Facilities inadequate	Planning, layout
2	Improper use of tools, equipment, facilities	Lack of skill and knowledge	Employee training
		Lack of proper procedures	Established operational procedures
		Lack of motivation	Enforcement of procedures Supervisor safety indoctrination Employee training and safety
3	Unsafe or defective equipment and facilities	Not recognized as unsafe	Supervisor safety indoctrination Employee training and safety consciousness
		Poor design/selection	Planning, layout, design Supervisor safety indoctrination Equipment, material, tools
		Poor maintenance	Maintenance and repair system
4	Lack of proper procedures	Omissions	Operational procedures
		Errors by designer	Planning, layout, design
		Errors by supervisor	Supervisory proficiency
5	Improvising unsafe procedures	Inadequate training	Established operational procedures Enforcement of proper procedures Employee training Employee safety consciousness
		Inadequate supervision	Supervisor safety indoctrination Employee training Employee safety consciousness
6	Failure to follow prescribed procedures	Need not emphasized	Enforcement proper procedures Supervisor safety indoctrination
		Procedures unclear	Operational procedures

Table 14.2 (Continued)

Immediate Causes	Possible Underlying Causes	Management Failures (Inadequacy)
7 Job not understood	Instructions complex	Operational procedures, planning, layout, design
	Inadequate comprehension	Employee selection and placement
8 Lack of awareness of hazards involved	Inadequate instruction	Supervisor safety indoctrination Employee training and safety consciousness
	Inadequate warnings	Planning, layout, design Safety rules, measures, equipment Operational procedures
9 Lack of proper tools, equipment, and facilities	Need not recognized	Planning, layout, design Supervisor safety indoctrination
	Inadequate supply	Equipment, materials, tools
	Deliberate act	Morale, discipline
10 Lack of guards and safety equipment	Need not recognized	Planning, layout, design Safety rules, measures, and equipment Supervisor safety indoctrination Employee safety consciousness
	Inadequate availability	Equipment, materials, tools Operational procedures
	Deliberate act	Morale, discipline, laziness
11 Lack of protective equipment	Need not recognized	Planning, layout, design Safety rules, measures, equipment Supervisor safety indoctrination Employee safety consciousness
	Inadequate availability	Equipment, personnel, tools Operational procedures
	Deliberate act	Morale, discipline
12 Exceeding prescribed limits, load, strength, speed, etc.	Warning inadequate	Safety rules
	Improper procedures	
	Inadequate instructions	Employee training
	Lack of comprehension	Employee selection and placement
	Deliberate act	Enforcement of safety rules Employee safety consciousness
13 Inattention: neglect of obvious safe practices	Lack of motivation	Enforcement of safety rules and measures Enforcement proper procedures Employee training Employee safety consciousness
	Inadequate comprehension	Employee selection, placement

Table 14.2 (Continued)

Immediate Causes	Possible Underlying Causes	Management Failures (Inadequacy)
14 Fatigue, reduced alertness, and hypnosis	Excessive physical or mental requirements	Planning, layout, design Employee selection, placement Operational procedures
15 Misconduct: deliberate failure to use protective clothing	Low morale and poor attitude	Supervisory training Employee selection, placement
	Misassignment	Planning, layout, design Employee selection, placement Employee training

Using the summary chart of Table 14.2, any of the basic causes can be tracked back to one or more general management failures or responsibilities. This process will answer the question "Where did management fail?"

SUMMARY

Once again we are turning a prevention tool into an investigation and analytical technique. The concept is not to simply detect and remove hazards but also to identify and prevent the recurrence of higher level management failures that allow hazards to exist and incidentally are probably permitting many other hazards to coexist.

The system departs from the principle that the immediate supervisor is the "key man" in mishap causation. It leads the way to direct and indirect involvement of higher management and staff functions.

In a way this is a systems approach. We are looking at the whole system, since the success of an operation is a management function and depends on the way that management allows their system to operate. This approach allows us to examine the operating system and the process to determine overall efficiency of the operation.

QUESTIONS FOR REVIEW AND DISCUSSION

1. Would you say the hypothesis you are asked to accept is true or false? Give reasons for your answer.
2. Comment on the definition of "management" given in this chapter.

3 Was it worthwhile to investigate such a simple mishap as described in the case "No Belt Guard"? Why or why not?

4 What other items might you add to the list of 15 basic mishap causes?

5 Give another example of poor housekeeping like the one shown in Table 14.1.

6 Supply a practical example for each of the immediate causes described in Table 14.2.

QUESTIONS FOR FURTHER STUDY

1 Is the National Safety Management Society still in existence? If so, write a paragraph describing its organization and status.

2 What kind of environmental factors could be included in Table 14.2? List at least five such factors in a format for adding to the table.

3 Could stress be entered in Table 14.2 as a management factor? Could lack of proficiency? Could failure to plan for computers on the workfloor? Give a concrete example of each by presenting sample entries for the three columns.

4 By reviewing your own files or talking to managers and safety personnel, find another case that can be used to illustrate the principles of the "Where did management fail?" approach. Write it up in much the same manner as the cases are presented in this chapter. Be ready to discuss the case and turn it in during class.

REFERENCES

1 Bruggink, Gerard M., "The Last Line of Defense," *Legal Eagle News*, March 1975, Vol. 17, No. 7, p. 6.

2 Grimaldi, John V., and Simonds, S. Rollin, *Safety Management*, 4th ed., Irwin, Homewood, IL, 1984.

3 Petersen, Dan, *Safety Management: A Human Approach*, Alroy, Englewood, NJ, 1975.

4 ———, *Techniques of Safety Management*, 2nd ed., McGraw-Hill, New York, 1978.

5 ———, *Human Error Reduction and Safety Management*, Garland STPM Press, New York, 1982.

6 Fine, William T., *A Management Approach in Accident Prevention*, Surface Naval Weapons Center, Silver Spring, MD, 1976.

CHAPTER FIFTEEN

Fire Investigation

Most mishaps with which we are concerned involve activities within structures or the grounds immediately surrounding those structures. Thus it is proper to center our attention on what are loosely and collectively called structural fires.

Two major areas of nonstructural fire investigation, (1) grass and forest fires and (2) transportation fires, are not detailed here, although many of the principles will apply. The transportation areas of rail, air, marine, highway, and pipeline have aspects peculiar to their operation that call for expert examination. The ideal situation is to have both a fire investigator and a subject matter expert working together on a fire-related investigation.

ARSON

A negative feeling about completing arson investigation has developed because of the many suspected arson fires that never make it through the courts to a successful prosecution. There are good reasons, however, for a positive attitude:

1 Many crimes are more difficult to detect than arson. Good routine fire investigation will determine if arson is probably present.
2 Arson evidence is generally as visible after a fire as any other type of evidence.
3 You do not have to see the arsonist light the blaze to convict him, as it is often said, but there are some special rules of evidence to consider.
4 The investigator can usually tell where the fire started. If arson is suspected, there are some special actions to take and investigative duties take on a wider scope. The arson investigator:
   (*a*) Must collect and preserve evidence that will be acceptable in court.
   (*b*) Will probably provide technical assistance to the prosecuting authorities.
   (*c*) Must properly interrogate witnesses. A suspect witness must be advised of his rights before an accusatory approach can be used. See Chapter 18 for legal procedures in investigation.

In many states and all Canadian provinces the primary responsibility for arson investigation rests with the fire marshal. If there is no fire marshall, the fire department or police department collects evidence to be given to the prosecuting attorney. Fires resulting in loss of life, large fires, and arson fires are also likely to be investigated by other parties, such as insurance companies, manufacturers of involved products, and local fire departments. A fire that suggests incendiarism should be reported by the fire department, and the local police and fire marshal's office should be called to assist in the investigation.

In fire investigation, whether arson is suspected or not, the investigator is interested in definite information:

1 The point of fire origin. If there is more than one point of origin, it is probably arson-related.
2 The source of ignition. Heavy burning along lower walls and floors signal an unnatural act, possibly arson. So does any sign of extreme heat, such as melted copper or spalled concrete flooring.
3 Why a small fire became big.
4 The reason for casualties. In fire deaths there should always be a forensic autopsy in case the fire is covering a homicide.
5 Why the fire deviated from normal fire behavior, if it did. This deviation is often a major clue in arson cases.

These five factors are common to all fire investigations, including arson investigations. A sixth, unlisted factor, beyond the scope of this chapter, is the need to find the extent of loss or business interruption.

When arson is deemed a possibility, some procedures are different:

1 The evidence requires a high degree of security for use in court. In the case of a non-arson fire the evidence may often be destroyed after the investigation is complete. The evidence is retained when arson is suspected.
2 The freedom of report writing and questioning is greatly curtailed once an arson suspect is identified, to protect the suspect's rights and to build a case.
3 When arson is established, it is usually not necessary to carry on the search for causal factors.

STRUCTURAL FIRES

One review of fires by source[1] shows six leading sources, listed below in order of frequency:

1 Trash burning.
2 Miscellaneous.
3 Suspicious incendiary.
4 Smoking related.
5 Electrical wiring.
6 Defective and misused heating and cooking equipment.

Item 2 is not much help to us, but it does indicate the magnitude of our problem in seeking better investigation. The other causes can be determined by the procedures that follow.

Ignition Sequence

To determine the ignition sequence we must determine where the fire sequence started. The ignition sequence has three factors: a heat source, a kindling fuel, and a human or natural event that combines the heat source with the starting fuel. Each of these factors must be identified separately to explain the ignition sequence. The fire ignition sequence is not always easily identified, but it starts with the point of origin. The ignition source of a fire can be traced by knowing the temperatures involved. Evidence of a 2000°F temperature (e.g., melted copper) tells us that burningl wood could not have directly resulted in that heat and that we must look further for the source. (See Appendix G for temperature guidelines.) To systematically move forward in our search for the source, we might consider the following guide, taken from the National Fire Protection Association's *Fire Protection Handbook*[2]:

1 Review the structural exterior, fire suppression, and timing.
2 Reconstruct as much as possible.
3 Find the approximate burning time and temperatures.
4 Find the path of heat travel and the point of origin.
5 Evaluate combustion characteristics of all materials involved.
6 Compare similar materials and situations.
7 Fit known facts to various possibilities.
8 Corroborate information from occupants and witnesses.

It is not a coincidence that the first item, the structural exterior, is where we start our serious investigation.

Outside

Many, perhaps most, serious fires occur during nonoperating hours when no one, not even a watchman, is on the premises. For such fires it is proper to start the investigation outside the structure. If we do not start here, faulty conclusions may be reached on the inside. Note whether doors and windows were open or unlocked. Noted whether the conditions was changed by firefighters during their effort. This is a good time to talk with eyewitnesses about what they have seen and to get their names for future contact. You should make a survey of the debris. Sometimes a laboratory analysis of the debris must be made to determine the condition or nature of the materials before a fire. Before any fire sequence evidence is taken from the scene, have it photographed and tagged or marked with the owner's name or initials. Mark the type of evidence it is supposed to be and the date and location of the fire.

Electrical[3,4]

Major buildings usually have controls and other key electrical equipment (switchgear, transformers, substations, and main switchboards) located in outside buildings, cutoff rooms, or vaults, or outdoors. It does not take long to find out if power was on or off during a fire. Neither does it take long to find if there have been major line disturbances, such as lightning.

1. If the power was definitely cut off at the outside, this knowledge will save much time inside the building.
2. If there was an electrical storm at the time of the fire, a lightning check definitely is in order. Lightning can enter buildings through supposedly dead lines. Check points of wiring entry for evidence of this.
3. As a logical extension of the outside check, check main internal controls first to ascertain the condition of switches, fuses, and breakers. If there is reason to suspect a wiring fault and the breakers have not been tripped, have the breakers checked by a qualified testing laboratory.

Flammable Liquids and Gases

It is common to find small to moderate amounts of flammable liquids in drums and flammable gas containers stored close to a building. Even more common are LPG (liquefied petroleum gas) dispensers. Where such facilities are known to exist, the presence or absence of windows, doors, and wall openings should be checked. Of particular importance is whether there were intake fans, especially those connected to air-handling systems, that could have drawn gases into the building. More than one fire has started when flammable liquids or gases from leaking containers entered a building and met an ignition source.

Inside

Before you can do a thorough investigation job inside, you should be reasonably familiar with fire behavior, common ignition causes, basic structures, and protection defects that might play a role in converting a small fire to a major one. You should be able to answer five questions that are common to arson investigation:

1 Where did the fire start?
2 What was the ignition source?
3 Why did a small fire become large?
4 What was the reason for any casualties?
5 Why did the fire pattern deviate from normal?

There are normally two lines of investigation following a major fire: first, tracing the fire and second, determining the loss. We will concentrate on tracing the fire. Before going deeper it may be well to discuss a major source of inside fires—smoking. With billions of cigarettes being lit daily it is not hard to see why smoking is a major cause of fires. Cigarettes continue to burn when they are left unattended or are discarded. They present a hazard when the smoker falls asleep or throws them into containers with combustibles. Careless smoking surpasses all other causes of fires in almost all types of structures. In one- and two-story dwellings it is the leading cause of fires.

Seek the point of origin with the following steps:

1 In the initial check, temporarily disregard the upper portions of the fire and concentrate on locating low burns.
2 For each low burn, note the location of the main fire by depth of char and other indications. Analyze the spread of fire away from each point.
3 Into the basic pattern of rising hot air, fit in the chimney effects of floor openings and shafts along with wind directions and various influences on the spread of fire.
4 Although there may be several low burn points, normally only one is the logical point for a fire to start. Minor low burns can be quickly identified as coming from a developing fire and not the initial source. If two or more burns appear to be of equal importance to the developing fire, the odds are that arson was involved in the fire. At this point an arson investigation would begin.
5 Assuming the fire started at a single low-point burn, the next thing is to determine the cause.

The ignition source could be many things: open flames, sparks, hot oil, steam pipes, motor bearings, and so on. For these sources to result in fire it is necessary for fuel to be present. The kind and amount of fuel is important in determining the ignition source. We will cover two principal ignition sources out of many, but some basic principles can be shown by the following examples of electrical sources and flammables.

Electrical

Electricity as an ignition source is a broad subject, and large books (Mazer,[4] for example) have been written just about that source. In reading the following section it might be well to keep in mind that electrical fires can be divided into two major classifications: (1) fire originating within the electrical distribution system, and (2) fires originating within electric equipment. Some fires involving electric appliances should not be thought of as electrical, such as when a child places paper in a toaster or leaves clothes too close to an electric heater.

Following a major fire, electrical wiring, conduit, fixtures, equipment, and the like are scattered about. The wiring proper seldom offers clues except to the major task of tracing a circuit. Nevertheless, there are usually some clues that can be checked out. One is the fusing of copper wire ends at a breaker or control point. This sign of internal power is discussed in more detail in Chapter 7.

A simple fusion at a control point, along with a blown fuse or tripped breaker, is sound evidence that a short circuit occurred and that the safety equipment was activated. The next point is to check if fuel was present that could have been ignited by the short circuit.

Multiple fusion points generally mean that the fire source has not been found. It is probable that other short circuits followed the initial short circuit. Possibly the breaker did not trip at all. This does not necessarily mean that the fire was not electrical in origin. The short circuits might have been caused by an external fire, and this should be checked out.

Although possible, it is rare for fire to be caused by a short circuit in conduit. Some conduit is galvanized, however, and zinc has a low melting point. Heat from an external fire can melt the zinc and cause a short circuit inside the conduit. This type of short circuit is the result of the fire and not one of its causes.

All electrical equipment is vulnerable to single phasing. Causes of single phasing on the usual three-phase system include a broken utility line, contact failure, wrong lead connection (new work or recent wiring), a grounded conductor, and other problems. Under such circumstances, equipment that has not been protected to compensate for single phasing overheats and becomes a potential fire source.

Many electrical fires are caused by the use of improper or incomplete equipment. There is often enough evidence left to check this out, particularly in fire-resistive buildings where much of the structure is left standing. Two simple illustrations follow:

1 If lightning is suspected, check the existence and adequacy of lightning protection. If it is missing and lightning did strike, the lightning's point of entry into the building can be plainly seen, as can part of its path inside.

2 Pendant lights on drop cords are widely used in place of the far safer pipe pendant lights. Where combustibles are present, a light bulb on a drop cord should have a wire guard and, in hazardous areas, a vaportight globe as well. There will be evidence of the guards and vaportight globes if they were present. Lack of a wire guard cost a government agency over $7 million a few years ago when a burning bulb in contact with combustibles was not turned off.

Flammable Liquids

Strictly speaking, flammable liquids do not cause fires, but many bad fires have resulted from their misuse, from faulty design, and from equipment failures. Among the most common sources of serious flammable liquid fires are the following:

1 Failure to control ignition sources is sometimes more important from a safety standpoint than from a property standpoint. For example, a small group of men were cleaning paint spots off the concrete floor of a newly painted building. After running out of the safety solvent, they siphoned gasoline from a car to complete the job. Gasoline vapors were ignited by the pilot light on a water heater. Four men died in the ensuing fire. Damage to the building was only $95. Similar fires have, of course, caused substantial property loss. Where misuse of flammable liquids is suspected, it is important to make sure that possible ignition sources for the vapors are not ignored. This is particularly important when vapors could have penetrated a lower story or basement.
2 Quench oil tanks are sometimes the source of serious fires. Such fires result when the oil is overheated for one reason or another or when hot metal particles get hung up on the passage through.
3 Poorly arranged piping for flammable liquid has contributed to bad fires. An example would be where a fitting has failed and flammable hydraulic oil then sprays directly into a furnace.

When the presence of flammable liquids is suspected, an investigator can quickly check for their presence by placing fire residue in a closed container of water. Most flammable liquids found in such a situation are not soluble in water and will float. A colored layer on top of the water shows their presence.

Investigators commonly find containers at a fire scene. A container's type, condition, and the presence or absence of flammables can tell a great deal. For example, it is reasonable to assume that a safety can that has bulged and split at the seams may have contributed to the fire but was not a primary source. A safety can or other sealed container that leaked would not likely build up enough internal pressure to split a seam.

Heating

Heating equipment is a common and sometimes complex source of trouble. Some methods used to find related fire sources are explored:

1. Check steam pipes to see if the pipes were too close to combustible materials such as walls or wood floors. Check vent pipes of stoves and other heaters. The metal stacks transmit fire to nearby sections.
2. Check dial records. It is common practice to automatically record various dial readings, to include those showing amounts of flammable gases in exhaust stacks, boiler water level, and pressure. These recordings often survive a boiler explosion and the ensuing fire and should be salvaged. The services of a boiler expert may be needed to properly interpret them.

Explosions

Occasionally the fire investigator will be asked to investigate explosions where no fire was involved. In such cases finding the origin and causes of the explosion may be all that is needed. More often, the explosion will result in a secondary fire. Then the fire damage may make it difficult to find the origin and cause of the explosion. Also, a fire may result in an explosion that is incidental to the fire, and its effects can then mislead the investigator. A short but good treatment of explosions possibly related to fire can be found in reference 5.

Spontaneous Ignition

Spontaneous ignition, formerly called spontaneous combustion, is the apparent ignition and burning of material in the absence of heat, spark, or flame. The term is misleading, and "self-heating" and "self-ignition" are sometimes substituted. Such ignitions are spontaneous only in the sense that flames suddenly appear on the surface of nonflaming material. Usually heat has built up over a long period—for several hours, days, months, or even years. There are a few exceptions to these lengthy periods; some reactive or pyrophoric material may produce flames almost at once when exposed to oxygen or another chemical.

There are four general types of spontaneous ignition:

1. **Biological Initiation.** The most common example of this occurs when hay has not been completely dried before bulk storage in barns. The moisture present allows biological action to take place. The heat released can eventually raise the temperature high enough to result in a fire.
2. **Chemical Reaction.** Another source is the oily rag. Unsaturated vegetable oils absorbed on cotton waste or rags can readily self-heat and ignite. In order of sensitivity, a few of these oils are: linseed, tung,

hemp, poppy seed, sunflower seed, soybean, corn, and cottonseed oils.

3 **Initial Heating of Materials.** Some materials, safe at room temperatures, can present hazards if stored at higher temperatures. Two examples are cellulosic fiberboard and glass fiber insulation.

4 **Pyrophoric Metals.** Some metals are pyrophoric in that they can ignite spontaneously in air or water under certain circumstances. These metals include plutonium, uranium, thorium, zirconium, magnesium, calcium, potassium, and sodium. The conditions that will produce ignition vary widely. For example, liquid sodium will react violently when dropped into water, and uranium scrap will ignite spontaneously under summer heat conditions unless stored under oil or water.

There are several test methods to determine whether a material may have undergone spontaneous ignition, but the services of a laboratory will probably be needed. The exact nature of self-heating has not been clearly defined, so expect varied opinions on the subject.

Miscellaneous

Miscellaneous sources of ignition number in the thousands. Some of these potential sources are missed because they are outside the usual experience of the investigator. For example:

1 When a fire appears to have started close to high-speed machinery, consider the possibility of friction as an ignition source. Underlubricated bearings on high-speed equipment can generate enough heat to ignite lint, flammable liquids, paper, and other materials. When friction is suspect it is worthwhile to check bearings of machines not deeply involved in the fire as well as past inspection reports to review the maintenance level.

2 Gas confined in piping is not a hazard but may become one when it leaks because of failure of pressure joints, line breaks or fractures, poorly sealed connections, poor valves, or other defects. If there is no source of ignition at the leak, it may go undetected until a gas–air mixture in the explosive range has formed. There will then be an explosion ahead of the fire. Failure of a gas line from heat during a fire results in a blowtorch effect at the opening and does not cause an explosion.

3 Severe fires have resulted from cutting and welding not done on a fire floor. Molten globules of metal can become combustible materials in several ways.

4 If a fire seems to have started in file rooms and cabinets, the investigator should see if it started above floor level. One of the first

things to do is to open the file drawers near the point of origin. By comparing the amount of damage in drawers it may be possible to pinpoint the drawer involved. It is then fairly easy to find out what was in the drawer before the fire and check out the possibilities of fire origin, for example, spontaneous heating, unauthorized materials, careless smoking, and so on.

Why Small Fires Become Large Fires

Knowing how and why small fires become large is as important as knowing why they start.

- **Location.** A few years ago virtually all major industrial and commercial buildings were located in heavily built-up areas with a well-staffed fire station nearby. Today many of these large-scale operations are located away from metropolitan areas. Unless they are large enough to have their own fire departments they are likely to be far from a well-equipped and well-staffed firefighting force. Response time is critical to fire control. In a metropolitan area we can expect response times of less than 5 minutes, while the average response time in the country is over 20 minutes. Even then they may be miles from a water source.
- **Delayed Discovery.** Many important buildings, including churches, schools, and warehouses, do not have watchmen, fire detection systems, or fire protection. Many of these structures are unattended at night and during many daylight hours, thus depending on causal passersby for fire detection. By the time it is reported, the fire has often grown to major proportions.
- **Delayed Alarms.** All too often, employees and others, on their own or by instruction, attempt to fight fires with the means at hand before calling the fire department. Many lives have been lost and untold millions of dollars worth of property damage has occurred because untrained or poorly trained individuals tried to fight the fire on their own.
- **Winds.** In the Southwest much property and some lives have been lost due to a stubborn refusal of property owners to recognize what wind can do in a dry country. The hot, dry Santa Ana winds come out of the desert, sometimes with speeds of 100 mph. Property owners take a chance when they plant combustible shrubbery close to wooden structures with shake roofs. The outcome is predictable when a fire starts: It sweeps through the housing area. Military installations and many industrial parks suffer little damage from this due to the separation of facilities.
- **Water Supplies.** The lack of water to fight fires speaks for itself, and reasons may range from short supply to freezing.

Reasons for Casualties

In the United States, where about 6000 people die annually from fires, most fire-related deaths do not result from burns. Even the recovery of a 100 percent burned body after a fire does not mean the person died from burns. A careful autopsy may prove death occurred before burning. Causes of such deaths include:

1. Carbon monoxide poisoning.
2. Carbon dioxide poisoning.
3. Oxygen depletion.
4. Excessive temperature.
5. Toxic gases from burning materials.
6. Smoke inhalation.
7. Murder.
8. Suicide.

The investigator does not usually investigate fire deaths, but he should make certain an autopsy is carried out to find the cause of death.

Unfortunately, most plastics are combustible, and some are very toxic when burning. Additionally, the manufacture of finished plastics uses large amounts of highly flammable materials. One summary[6] of other studies looks into 75 plastics commonly used in airlines alone. This one report gives us an idea of the problem size.

Wiring Defects and Control Devices

Wiring defects are the most important type of electrical defect, which in turn ranks first in fire causes. Ninety percent of all wiring failures and related fires can be traced to the following list:

1. Warm and hot fuse boxes, meaning overloaded circuits.
2. Repeated tripping of circuit breakers
3. Oil-soaked wiring, particularly on the floor or in floor trenches.
4. Ordinary wiring in hazardous locations, such as acid rooms.
5. Tandem cords and multiple outlets.
6. Extension cords of lower amperage rating than the appliance.
7. Overlong extension cords.
8. Splices and taps outside approved conduit boxes.
9. Wiring on floors or under carpets.
10. Drop wiring from ceiling fixtures not approved for the purpose.
11. Portable lights without fire guards.

12 Circuit breakers with contacts taped closed.
13 Wiring fastened to walls with nails or similar fasteners.
14 Jumpers across cartridge fuses, wrong size fuses, or metal disks or coins behind fuse plugs.
15 Loose fittings and supports in vibration areas.
16 Excessive temporary wiring.
17 Improperly grounded circuits.

Some of the more common defects related to control devices that lead to fire are:

1. **Overload.** Frequent blowing of fuses or repeated operation of circuit breakers is a clear sign of a possible overload situation. Warm fuse boxes are a similar indication.
2. **Jumping.** This is the bypassing of fuses so that they will not have to be replaced. Coins behind fuses and copper strips for cartridge fuses are the usual short-circuiting methods.
3. **Overfusing.** This refers to using fuses larger than required by the wiring to keep from blowing fuses.
4. **Breaker Tiedown or Tieback.** Breakers are tied so they cannot function properly.

Fire Behavior and Fire Patterns

In order to work back from a major fire to a point of origin, the investigator must have at least a basic knowledge of fire behavior and fire patterns. The following discussion emphasizes structural situations, but much of it applies to nonstructural situations. For a more complete picture, refer also to the material in Chapter 7, Systems Investigation.

Structural Fire Behavior

1. Hot gases including flames are lighter than air and therefore rise. There must be a physical obstruction or strong air currents for the fire to burn in any other direction. Without those forces the fire will always burn upwards.
2. Combustibles in the path of rising flames ignite and increase the fire volume and the rate of upward movement. If this takes place in a restricted passage the velocity increases as a "chimney effect" is created.
3. A major fire can develop only when fuel is above the initial flame. Exceptions are rare, being noted in only 2 percent of the cases.
4. Strong wind currents can modify this upward reach of flames:

(*a*) Natural ventilation or deflection may have strong lateral or downwind currents.
(*b*) Ventilating, air-conditioning, or even escalator systems can induce a downward movement of flames.
(*c*) Lateral fire spreads rapidly when there is a physical barrier to vertical movement and combustibles to feed on.

5 Downward spread occurs rapidly when highly flammable coatings or surface treatments are present.

Fire Patterns

We can draw some conclusions based on the structural fire behavior described above:

1 Every fire forms a pattern that tells a story.
2 A fire following a normal pattern develops as an inverted cone with the apex at the bottom. In a multistoried building, a new inverted cone may develop for each fire breakthrough from below.
3 When the fire pattern on the floor of origin shows two or more inverted cones, there is every reason to suspect arson.
4 The point of origin does not have to be a physical point, although the origins of Class A and C fires come close to that. The point of origins of Class B fires is flattened into a smaller area, and the size depends on the amount of free flammable liquid available.

Collateral information combined with the above information makes the job of tracing a fire to the point of origin less complicated than it might at first appear.

FIRE INVESTIGATION PHOTOGRAPHY

Photography is a useful tool for the fire investigator to document visual observations, emphasize fire development characteristics, and confirm physical evidence found at the fire scene. The usual purpose is to produce useful and legally acceptable photographs for reports and courtroom presentations. Most of the guidelines given in Chapter 3 apply. This section discusses only the fire investigation aspects.

Fire investigation photographers have a systematic procedure to follow:

1 **Exterior Photographs.** The first step is to document the exterior of a structure, vehicle, or object before probing the cause of the fire.

2 **Interior Photographs.** The second step is to document interior damage, showing the progress and extent of the fire, including room(s), area(s), and point(s) of fire origin. These pictures are made before digging or probing the fire debris and will document conditions on the investigator's arrival.
3 **Interior Photographs, Postclearance.** This step concentrates on the debris-cleaning operations, char and burn patterns, and the evidence before its removal from the fire scene. Should associated crimes, such as breaking and entering, theft, or homicide have taken place with the fire, any physical evidence of these acts should also be photographed.
4 **Mosaic Photographs.** If the photographer is familiar enough with his photo equipment, he might try a series of overlapping photographs. Later the prints can be put together to show large areas such as inside a warehouse.

It is extremely important to fully document each photograph and carefully control its handling through processing and later use. For later courtroom use the chain of custody is a factor, as explained in Chapter 18.

SMOKE AND FLAME

It is common for the fire photographer to arrive at the site of the fire while the fire is in progress. This allows an opportunity to document the actual progress of the fire. Color photographs can show several things about the fire, one being the color of the smoke and flames, which are indicators of the types of materials burning. Examples are given in Table 15.1.

Caution is required since the investigator/photographer may arrive during the later stages of the fire. Most structures contain fuels with hydrocarbon

Table 15.1 Smoke and Flame Indicators of Burning Material

Fuel	Color of Smoke	Color of Flame
Wood	Gray to brown	Yellow to red
Paper	Gray to brown	Yellow to red
Cloth	Gray to brown	Yellow to red
Gasoline	Black	Yellow to white
Naphtha	Brown to black	Straw to white
Benzene	White to gray	Yellow to white
Lubricating oil	Black	Yellow to white
Lacquer thinner	Brownish black	Yellow to red
Turpentine	Black to brown	Yellow to white
Acetone	Black	Blue
Cooking oil	Brown	Yellow
Kerosene	Black	Yellow

bases and, when burning, these fuels may produce smoke and flame that could mislead the investigator.

The color of the flame is also related to temperature of the fire, for example:

1500°F	Cherry red flame
1830°F	Light red flame
2000°F	Orange flame
2200°F	Yellow flame
2400°F	Yellow to white flame
2700°F	Dazzling white flame

ADDITIONAL SOURCES

Several private, state, and federal organizations are engaged in activities that will lead to more effective fire investigations. The federal ones are listed below:

Bureau of Alcohol, Tobacco, and Firearms
1200 Pennsylvania Avenue, N.W.
Washington, DC 20226
  Explosives Enforcement Branch
  Explosives Technology Branch
  Forensic Branch

Federal Bureau of Investigation
9th and Pennsylvania Avenue
Washington, DC 20535

Law Enforcement Assistance Administration
633 Indiana Avenue, N.W.
Washington, DC 20531
  National Institute of Law Enforcement & Criminal Justice

National Bureau of Standards
Washington, DC 20234
  Center for Fire Research
  Law Enforcement Standards Laboratory

National Fire Academy
U.S. Fire Administration
Route 1, Box 10A
Emmitsburg, MD 21727

National Transportation Safety Board
800 Independence Avenue, N.W.
Washington, DC 20594

U.S. Fire Administration
2400 Union Street, N.W.
Washington, DC 20472
Office of Planning & Education

U.S. Forest Service
P.O. Box 2417
Washington, DC 20013

SUMMARY

There are several specialties within the fire investigation field that are not addressed here. The common area of structural fire investigation is stressed, but there are specialists for nearly all transportation modes, forest fires, high-rise fires, school fires, fires in recordkeeping areas, film fires, and so on. There are also many associated specialists to call on, such as fire photographers, smoke specialists, laboratories that specialize in fire investigation, and engine fire specialists. We have here enough background to proceed with the average industrial or business fire situation and to know when to call an expert or the authorities.

QUESTIONS FOR REVIEW AND DISCUSSION

1 Who investigates a suspected arson case where you work?

2 What is the difference between the "source of ignition" and the "point of origin"?

3 Explain why a forensic autopsy should be conducted when a body is found in the fire debris.

4 If firefighters responding to a fire found the main outside electrical switch turned off, what could we assume about the fire?

5 If we spilled a puddle of gasoline on a clean tile floor of light color and immediately lit it, would it leave a stain? If so, what shape would the stain be?

6 If several exposed strands of copper wire are exposed to a 1500°F flame, what will happen to the copper strands? What will be the color of the strands? Will they melt? Will they fuse together? If you cut one open, will it be bright inside?

7 What will happen if we drop a match into a dish of gasoline?

8 Would regular car gasoline (unleaded) burn at a lower temperature than jet fuel?

9 Is there an organization of professional arson investigators? If there is, what is it called?

10 What are the four steps of the recommended systematic procedure for fire investigation photography?

11 Driving through the industrial section of town you see a fire. The fire department has not yet responded. You see thick black smoke and yellow to white flames. What can you assume about the nature of the burning material and the temperatures involved?

12 List three types of spontaneous ignition. Give an example for each.

QUESTIONS FOR FURTHER STUDY

1 By placing a telephone call or making a personal visit see what an arson investigator has to say about the adequacy of a regular accident fire investigation when arson is suspected. Write a summary of the comments.

2 If you drop a cigarette on a wooden platform, will the platform catch fire? Would a piece of paper catch fire?

3 Draw a typical flame, soot, and smoke pattern that we would see if the side of a camper caught on fire as it was driving down the highway at 65 mph.

4 Four steps are recommended for a systematic approach to fire investigation photography. Bring an example of each type of photograph to class. A good photocopy will suffice.

REFERENCES

1 Bugbee, Percy, *Principles of Fire Protection,* National Fire Protection Association, Boston, 1978, p. 127.

2 *Fire Protection Handbook,* 16th ed., National Fire Protection Association, Boston, 1986.

3 Whitman, Lawrence E., *Fire Prevention,* Nelson Hall, Chicago, 1979.

4 Mazer, William M., *Electrical Accident Investigation Handbook,* Electrodata, Inc., Glen Echo, MD, 1982.

5 *Fire Investigation Handbook,* NBS Handbook 134, Department of Commerce, Washington DC, 1980.

6 Graham, Lauren, "Research into Postcrash Fires" *Aviation Engineering and Maintenance Magazine,* Hamilton Burr Publishing, Santa Clara, CA, October 1977.

PART FOUR

Final Touches

The four chapters that follow tie together all that has gone before and speculate on future developments. The most important part of the investigative process, the mishap report, is discussed in Chapter 16. Reporting is equally important for the simplest of mishap and the catastrophe. Chapter 17 overviews the mishap investigation process with two approaches to a complete investigation. The first provides a 12 step sequence to make certain that one is ready to investigate, and more important, what to do after the hands-on part of the investigation is complete. The second approach provides a package that assures both a thorough investigation and steps for implementing the findings into mishap prevention. The legal aspects of investigation, which have recently received a good deal of public attention, are treated in Chapter 18. The reader will note that some of the concepts advanced there conflict with material given earlier in the book. That is the way it will always be as court decisions change the legal process. We have been looking at investigation for prevention purposes, but the process is handled somewhat differently for legal purposes. In fact, some professional investigation experts do all their work with the legal system in mind and never have a thought for prevention. Finally, in Chapter 19 we look to the future of mishap investigation. Covered there are new techniques, not fully proven, and the future roles of mishap investigation in government, business, and industry.

CHAPTER SIXTEEN

The Mishap Report

The best investigation flounders in obscurity unless the results are presented clearly and concisely in a report. Written factual conclusions and recommendations should make clear the need for corrective actions. Before reaching the party responsible for corrective action someone must, on the basis of the report, make specific corrective actions needed to prevent similar events from happening again. The value of an investigation, even one incorporating the appropriate techniques discussed in the earlier chapters, rests on how well the report is written.

REPORTING REQUIREMENTS

The OSHA Act of 1970 requires that if a mishap results in a fatality or the hospitalization of five or more employees, the employer must report the event orally or in writing to the nearest OSHA office. No particular format is prescribed, but the report should include an explanation of the circumstances related to the mishap, the number of fatalities, and the extent of any injuries. This allows the office to dispatch a compliance officer to the scene if a mishap investigation is necessary. The purpose of that investigation is to determine if a violation of any standard may have contributed to the mishap. This is not well aligned with the concept that considers an investigation to be made for purposes of future mishap prevention.

Some states have their own OSHA-type requirements. Since their requirements need at least as much information as the federal OSHA, we can expect their rules to be more stringent. For example, California requires a report to their nearest CalOSHA office within 24 hours of an event requiring hospitalization of one or more persons for over 24 hours for other than observation; one that resulted in disfigurement or dismemberment; or one in which five workers are involved in the same incident or death. The employer's first report of a mishap, injury, or incident is to be completed and submitted to the insurance carrier. Deaths from a vehicle collision do not normally require an OSHA report.

Many state codes also require that an employer file with the insurance carrier a report of every injury or illness arising out of the course of employment, and include detailed reporting requirements setting conditions and time limits concerning the filing of the report. All insurance carriers have similar requirements to ensure that they are notified. We can expect that the efficient investigation conducted as directed in this book will contain far more data than are needed to meet regulations.

Some agencies may have reporting requirements for mishaps that fall within their area of responsibility. This is illustrated in boating mishaps. The Coast Guard have their own boating accident report form,[1] which is Coast Guard Form CG-3865. This form must be completed and submitted by the boat operator when there is (*a*) a loss of life, (*b*) injury causing any person to be incapacitated for more than 72 hours, or (*c*) actual physical damage to property (including vessels) over $1000. Five days is the normal reporting time unless loss of life is involved, in which case the report must be sent in within 48 hours. Many agencies have their own requirements, and they may conduct their own investigation of the mishap. While such reports may fill the needs of the agency, they seldom fill the mishap prevention need of the employer, and a complete investigation should be carried out within the organization.

This duplication sometimes results in several investigations of the same mishap. The collision of two aircraft over the New York City area several years ago illustrates the point. One aircraft fell into the harbor waters of New York, while another fell into a military airfield in New Jersey. In theory, and sometimes in practice, the result would be as follows:

1. A Coast Guard investigation was performed, since an aircraft fell into waters for which they have responsibility. They were also in charge of the aircraft recovery.
2. The NTSB investigated, since civil aircraft and fatalities were involved.
3. The FAA investigated since the air traffic control system was suspect.
4. A military investigation took place, since wreckage fell on a military airfield and military personnel were involved in the post-disaster activities.
5. An investigation was conducted by the Port Authority of New York and New Jersey because they are mandated to investigate mishaps that occur in their territory.
6. The State of New York had a similar mandate and investigated the mishap.
7. The Airplane Pilots Association was eager to keep their pilots from being blamed unfairly and conducted an investigation.
8. The airline involved had their safety representative conduct a separate investigation and make a report.

In today's world we could also expect representatives of the air traffic controllers union to make their own investigation to protect the interests of their members. We could also add to this list the trend for local district attorney's offices to conduct their own investigations in cases where negligence is an apparent causal factor in injuries or deaths.

This exaggerated example shows the stakeholders and the need for information. It is unlikely that any one person or group saw the resultant stack of reports, but their number leads us to believe that the process can be overdone. In practice, there are several ways the interests of the stakeholders can be protected (by inviting them to participate in one large investigation). In catastrophes this is often the case.

The Occupational Safety and Health Act provides for specific reporting requirements relating to mishaps,[2] which can help build mishap recordkeeping files. They entail five steps:

1. Obtain a report on every injury requiring medical treatment. The format of the report is not specified.
2. Record each injury on OSHA Form No. 1. Specific instructions are in OSHA guidelines.
3. Prepare a supplementary record of occupational injuries and illnesses on recordable cases on OSHA Form No. 101 or on worker's compensation reports requiring the same information.
4. Prepare an annual survey of mishaps and illnesses (OSHA Form No. 102). In California this takes the form of maintaining a Log and Summary of Occupational Injuries and Illnesses on the premises and posting it therein during the month of February.
5. Maintain the records in your files for 5 years.

INFORMATION SYSTEMS

Good investigations with solid, specific recommendations for corrective action can be of value long after the mishap if the investigation results are available. Similar mishaps may follow and other parties will have mishaps, but the information is useless unless it can be made widely available and readily retrieved.

Having information from a previous investigation at hand if a similar event happens gives us direction and can point out the effectiveness of our corrective actions. Summaries of mishap findings can show areas that need special emphasis. We may believe that we can recall the details of mishaps that occur in our own company. But consider the manufacturer with a high mishap rate—with 6000 employees at one location there were 370 recordable mishaps in one year. That manufacturer's managers cannot possibly recall all the important data from these mishaps. In fact, they process the data by hand. In the event of a severe back injury, for example, they can only see how many they have had by manually counting through the records. They clearly need a mishap information system just to know how they are doing and to find out where their problems lie.

For the small company with a recordable mishap a couple of times a year, the problem is not so acute. But their prevention effort could be better directed if they had access to the experience of other, similar businesses working under similar conditions. Once again the need for a mishap information system is seen, this time among several companies.

We might think that OSHA or some other mishap collection system will fill our need. Unfortunately the reporting requirements are weak. They often require only minimal OSHA recordkeeping information or records for insurance uses and are not at all suitable for prevention activities. Even the reporting forms used often lend themselves to little more than identifying an area of concern. Although this is better than nothing, much more specific information is needed to prevent mishaps.

Nevertheless, several mishap information systems are in effect. Some are highly computerized and efficient, such as those used by the military. Unfortunately, the average business cannot access these specialized systems. Many companies have computerized mishap data systems, but many suffer from poor design and inadequate investigative reports. Such systems are becoming more commonplace and more efficient.

Typical of the better systems is that used by Gulf Canada, Ltd. The system, called LOMIS,[3] is designed as an inquiry system to provide information and analysis of all reported mishaps. It is meant to increase research, to improve management's ability to diagnose specific problems, to assist in setting priorities, and to identify the most cost-effective remedies. This system is fairly advanced as far as general industry use is concerned today, but since input data are on a simple one-page coded report that can be filled out in a few minutes, it causes us to suspect the thoroughness and validity of even this system. For the investigative viewpoint LOMIS has value as a predictive tool by which management is briefed on significant downgrading mishaps, problem areas and locations, faulty systems and equipment, procedural weaknesses, training deficiencies, and unacceptable trends. It thus becomes of magnified value to management because it can provide meaningful financial returns and built-in justification for thorough investigation. Even with its limits the system increases the effectiveness of good investigation severalfold by making the digested information available in meaningful, directed terms. This particular system was a quantum jump over a noncomputerized information system and ranks high among those now in existence. Fortunately we can look forward to wider use and even better design of such systems.

One interesting use of computerized mishap information is found in some government agencies. When the agency is notified of a mishap and it is decided that an investigation is needed, all facts known about the mishap are fed into the computer. The computer, in effect, is asked: "Where should we focus our attention?" After an analysis of findings in its data bank about similar mishaps happening with similar equipment under like circumstances, the computer suggests where to focus attention by displaying and printing out what causal factors were present in the documented cases and where processes broke down before. Since we seldom find new causes[4] for our mishaps, this computerized guidance stands a good chance of being effective. Good historical data and analysis can be of great assistance, particularly if all we must do is query a computer to get a readout to take to the mishap scene with us. It is clear, however, that the information we get out of the computer can be no better than what was entered. This makes another clear case for efficient investigations and detailed, in-depth encoding.

INFORMATION COLLECTING AND REPORTING PROCESS

The five-step reporting process (facts, analysis, conclusions, recommendations, and summary) illustrated by Johnson[5] can apply to an investigation effort of any size. It is described below with refinements and guidance.

Facts

Facts should be presented in the report in a logical sequence, stressing those that bear on the mishap process and causes. Factual information carries the most weight. However, information that seems to be factual but cannot be proved may also be included. Supporting data carry the weight of fact when in enough quantity and if properly evaluated. This is the place to eliminate the unsupported hypotheses and uncertainties.

Analysis

In this phase of the report, all the factual information gathered is analyzed. The facts, conditions, circumstances, and inferences are weighted to provide a basis for the conclusions that follow. Information is not added during this phase. This is the time to order and analyze the facts.

Conclusions

The conclusions are based on what is known and what is not known after the available information and facts are analyzed. Since we subscribe to the concept that all causes, inefficiencies, and deficiencies found that pertain to the mishap are important, it is best that conclusions not be listed in any priority. If the investigator can remember that it is the process, the system, and the operation that are being critiqued, then all causal factors will be important. Do not include conclusions that cannot be supported by the analysis step.

Recommendations

The development of recommendations is so important that it is dealt with later in a separate section. Recommendations are the reason for the entire investigative process. Specific corrective actions cannot be taken unless there are specific recommendations to act on. Recommendations should not be combined. Each should indicate one particular thing to be done. This allows the assignment of individual corrective actions to responsible and accountable parties. It assures corrective action directed at a particular area. We can expect several specific corrective recommendations to result from even a simple, uncomplicated investigation of a minor event. Single, specific corrective actions are much easier to comply with.

It does not always follow that the investigator is the one who makes the recommendations for corrective action. If the analysis of information, facts, and conclusions is properly carried out, a skilled analyst can prepare single specific recommendations. It is best for the investigator to make recommendations, but some of the foremost investigators are neither interested in or have the skill to make solid recommendations for corrective action. We cannot, however, consider the investigation complete until specific corrective recommendations are made in a manner that allows timely and orderly follow-up action. Most safety professionals will not close out the process until corrective actions have been completed.

Mishap Summary

The summary is compiled last, but it should be placed first in the report so that the reviewers and readers unfamiliar with the event can derive essential information quickly without searching through the entire report. This is only good staff work by the report preparer. The summary should be a brief account of critical information about the event, giving the who, what, when, and where of the mishap. Protocol of the reporting organization may direct that the summary be placed late or last in the report.

A Refinement

The previously described five-step process has been refined a bit by Department of Energy investigators[6] with the addition of analysis steps before the Facts and after the Conclusions steps. This seven-step process is listed below and depicted in Figure 16.1.

1 Analyze all available information.
2 Isolate relevant and irrelevant facts. Develop hypotheses to resolve uncertainties.
3 Analyze the facts developed to date.

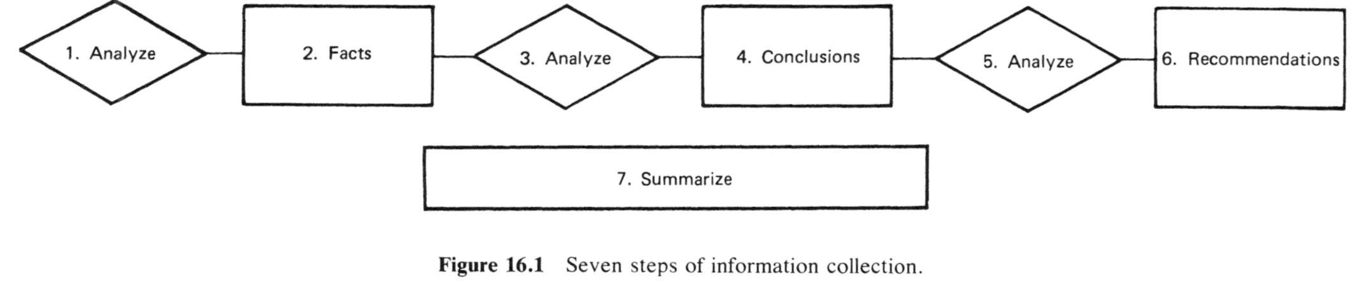

Figure 16.1 Seven steps of information collection.

4 Form conclusions based on what is known and what is not known, and determine serious deficiencies in information.
5 Analyze the conclusions for validity.
6 Make recommendations based on an analysis of the conclusions.
7 Summarize the entire process.

An Elaboration

Wood[7] has suggested that analyzing factual information should be a process of looking at each event's or operator's (actor's) phase of operation by examining it for facts, analyzing the facts, and reaching conclusions on each item. At this stage these "miniconclusions" may be positive or negative—that is, this action or actor did or did not contribute to the mishap. The positive conclusions are then listed chronologically as a major step in reaching a final conclusion. This logical extension of steps 1, 2, and 3 previously listed in "A Refinement" is graphically shown in Figure 16.2. Only three items are examined in step 2 and the figure, but many times that number may be looked at in a single mishap.

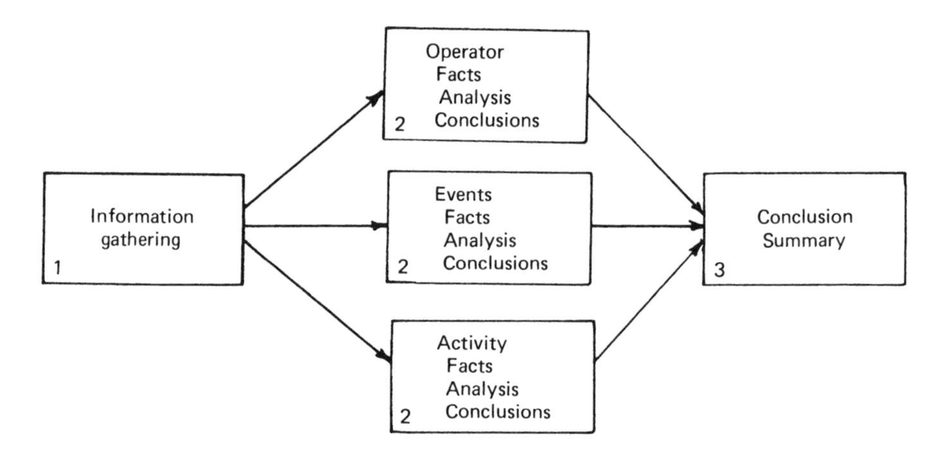

Figure 16.2 Examination of items of consideration through a facts, analysis, conclusions, and process.

Getting It All Together

Report preparation starts in a modest way at the beginning of the investigation. The apportionment of effort among investigation, analysis, and reporting is shown by Figure 16.3. The following steps are suggested for putting the report together:

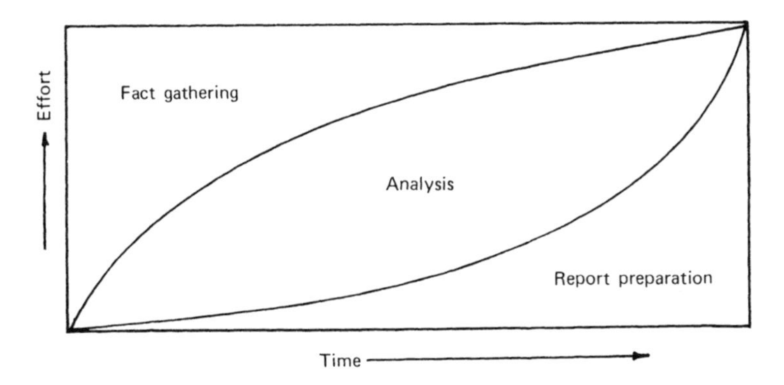

Figure 16.3 Resource allocation in mishap investigation.

1. Give a brief two- or three-line synopsis of what happened without investigative actions or conclusions.
2. Present a chronological history of the mishap, but without investigative details.
3. Select the facts, analysis, conclusions, and recommendations to be included, and present a structured picture of the investigation from fact gathering to conclusions and recommendations.
4. Select and design graphics to support and simplify the report. Graphics should simplify the text and should not be used unless needed to explain the text. They should not be used as filler. If not referenced in the text, they do not warrant inclusion in the report.
5. Select necessary backup materials and assign these to appendixes or file. The rules for graphics apply. If they are not critical to understanding the report, leave them out.

This book advocates "lean and mean" reporting. All information should be included that allows the reader to see exactly what has happened and take corrective action. Not one word more is needed, and not one less should be permitted. Good report writing is as much an art as mishap investigation and requires the same careful attention. Where thorough investigations are carried out with good recommendations for specific corrective actions finalized, too often the fine effort is hidden by excess information and appendixes.

Copies

How many copies of the report should be made? As few as possible. We can learn a lesson from the military, where even sophisticated reports are often restricted to three copies: the original and two more. Those three copies serve all purposes of a complete chain of command, a review process covering several persons or groups, a prevention requirement, and other purposes (liability, litigation, line of duty finding, and so on). Spare copies are not kept on hand. If needed, extra copies are made by the responsible office. This admirable trait of making only a few copies is, at this revision, sinking into obscurity. When safety directors are questioned, they are often found to be making many more copies of reports than needed, frequently 10 to 15. The reasons for so many copies vary.

Too often the investigator is so proud (and often rightfully so) of his work that a copy of the report is sent to every supervisor, every valued associate, and sometimes to family members. Distribution of 40 or 50 copies is not all that unusual, but only two or three are needed.

When information is needed for a certain reason, such as prevention or public relations, an extract will suffice. When persons and operations of interest to everyone in the organization are involved, nearly everyone would like to have their copy of the report. This is not valid.

The report is a desirable item for use in litigation or to assess blame and decide liability. The reports, now more than ever, figure prominently in court cases. This in itself is a good reason to keep the number of copies to a minimum. Wide circulation of reports increases exposure to unwarranted court actions. This, unfortunately, inhibits free exchange of information found in the investigation.

Reports should be circulated on a need-to-know basis and not to curiosity seekers. Access on a "freedom of information" or mishap prevention basis is handled through the report office of record.

DEVELOPING RECOMMENDATIONS

Major problems in developing recommendations have been summarized by Wood,[8] who has placed the role of the investigator in perspective for this purpose. A first concern is whether the investigator should even develop recommendations. Like so many answers, it depends on the circumstances.

The investigator should properly identify the problem areas, that is, what needs to be fixed. The need for expert investigation is widely recognized, and we have built corps of experts in nearly all countries. If recommendations are so important, why do we not also develop expert recommenders, persons fully trained and with appropriate background for the task? There is merit to this suggestion, but without likely early action a better job can be done with current resources.

The experience of investigators should not be ignored. They are usually able to make valid recommendations. The preventive action must start somewhere, so why not with them? We are intolerant of persons who bring us problems without solutions. Unfortunately, many investigators are trained only to find the problems. Some of the most highly skilled and specialized investigators fall into that category.

The investigator often forms opinions during an investigation that, if acted on, would increase the level of safety. It is his place to make these recommendations. If we accept them as being sincere and informed, but not necessarily practical and feasible, then that is one thing. If, however, the recommendations are "chiseled in stone," as most of the report is (since we cannot go back and conduct another investigation), then we have a serious problem.

Recommendations must be feasible. This is determined by cost, time, parts posture, tools availability, equipment on hand, weight or strength of the change, accessibility of the part to be changed, whether the change can be maintained, possible side or ripple effects, etc.

The true expert may not wear both the investigative and preventive hats, but in practice the investigator may have to. In most industrial concerns investigations are made by the first-line supervisor or the safety person whose main duty is prevention. The first-line supervisor also usually has the job of taking corrective action at the working levels. We will probably continue with the situation where the investigator makes recommendations that become the basis for early corrective and preventive actions.

Those recommendations should be viewed in the proper perspective. The office receiving the suggestion always has the option of accepting or rejecting them (with justification) or of finding an alternative solution that affords an equal or better degree of safety. There should be no pressure to adopt a recommendation if a better solution exists.

REPORT PREPARATION HINTS

Over the years mishap report reviewers have offered many ideas for improvement. Not all suggestions will apply to your own situation, but there is something for everyone in the items that follow:

1. Include the date of the mishap on the cover page.
2. Do not use proper names on the cover or in the report title.
3. Write the summary after the rest of the report is complete.
4. Back up the summary with information in the body of the report.
5. Do not use abbreviations or acronyms in the summary. It should be self-explanatory.
6. Do not refer to the body of the report or to figures and appendixes in the summary. It should be self-contained.
7. Avoid capitalizing job titles to increase readability.
8. Watch verb tenses, particularly between "what is" and "what was."
9. If you are using the events and causal factors charting in Chapter 11, a simple chart placed right after the summary can enhance understanding.

10 In drawing a map, include the area of focus without much other detail.
11 Indicate any use of medical personnel. If there is an injury, and medical personnel were not used, indicate why they were not.
12 Make certain there are good bridges from "facts" to "conclusions" to "recommendations."
13 Avoid using jargon unless it is needed to understand the mishap.
14 Do not place facts in the appendixes. Place them in the facts section.
15 Do not place conclusions in the facts section.
16 Use the necessary indices and scales in the graphics.
17 Write on the investigation each day. Don't let it pile up.

SUMMARY

The investigation report action really starts before the mishap occurs, when reporting requirements are known and made a part of the premishap preparation. Once the investigation starts, the report preparation starts with the gathering of facts and recording of the investigative and analytical process. As conclusions are reached and recommendations for corrective actions are made, the reporting effort reaches its peak, ending in a package calling for corrective actions.

QUESTIONS FOR REVIEW AND DISCUSSION

1 If there is a serious mishap where you work, who in particular must be notified under the federal or state requirements?

2 Can you think of anyone else who must investigate the aircraft mishap discussed under "Reporting Requirements" at the beginning of this chapter?

3 What do we do about OSHA reporting requirements in a state with a federally approved OSHA plan?

4 Do you use the regular mishap reporting form to report a fire within your operation?

5 Bring in a copy of the reporting form required to report a fire in your workplace.

6 Is a conclusion the same as a causal factor or finding?

7 Write at least three specific recommendations for corrective action based on the last mishap that directly involved you.

8 What is the difference between a synopsis and a summary in a mishap report?

9 How many copies of a mishap report must be completed where you work? Exactly who gets the reports?

QUESTIONS FOR FURTHER STUDY

1 If an employee falls off a ladder and breaks his leg falling against a steam kettle, do we have to notify the fire marshall because he is bady burned?

2 What nearby government agency has computerized information regarding mishap information? Secure a copy of their investigation reporting form.

3 What does LOMIS stand for?

REFERENCES

1 *U.S. Coast Guard Boating Guide,* Pocket Books, New York, 1975.

2 *OSHA Handbook for Small Business,* U.S. Department of Labor, Washington, DC, 1977.

3 *LOMIS,* Gulf Canada, Ltd., Toronto, 1979.

4 Ferry, Ted S., "No New Causes," *Directions in Safety,* Charles C. Thomas, Springfield, IL, 1976, pp. 215–222.

5 Johnson, W. G., *MORT Safety Assurance Systems,* National Safety Council, Chicago, and Marcel Dekker, New York, 1980, pp. 347–379.

6 Buys, Richard, et al., *Investigation Seminar, Las Vegas, NV,* EG&G, Idaho, December 1979, pp. 5-1 to 5-4.

7 Wood, Richard, Accident Investigation School, University of Southern California, Los Angeles, February 1980.

8 Wood, Richard H., "How Does the Investigator Develop Recommendations?", *Proceedings,* Annual Seminar of the Society of Air Safety Investigators, ISASI, Montreal, September 1979.

CHAPTER SEVENTEEN

Management Overview and Mishap Investigation

If benefits are to accrue to the company, investigation must consist of considerably more than completing the few actions required by the government. Our concern should be that every mishap investigation is a good one, thoroughly and properly conducted. It becomes the role of the safety professional or the experienced investigator to educate management as to the value of complete mishap investigation.

However, supervisors and untrained committee members conduct most mishap investigations. How, then, can management be assured that the best possible job is being done? How can they prepare these people to conduct good investigations, and how can they check them before, during, and after an investigation? There are three keys to successful investigation:

1. Be thoroughly prepared for investigation before a mishap.
2. Know how to gather and analyze the facts surrounding a mishap.
3. Have a good mishap report that serves as a basis for corrective action and process evaluation.

TWELVE STEPS TO A THOROUGH INVESTIGATION

One major problem is that there never seems to be enough time to make a good investigation. Regardless of the available resources, however, close consideration of the following 12 steps can ensure a quality effort:

1. Understand the need for investigation.
2. Prepare for an investigation.
3. Gather the facts about the mishap.
4. Analyze the facts.
5. Develop conclusions.
6. Analyze conclusions.
7. Make a report.
8. Make appropriate recommendations.

9 Correct the situation.
10 Follow through on recommendations.
11 Critique the investigation.
12 Double-check the corrective action.

Illustrating these as ascending stairs (Figure 17.1) helps to visualize their proper order. It will be seen that only three or four of the steps apply to the actual digging out of facts. Steps 3–6 are clearly in the domain of the investigator. Steps 1, 2, and 8–12, however, are not so clear. Step 7 is a gray area. In fact, the supervisor or foreman generally fills out the mishap report form, a task that may be performed in the most routine and perfunctory manner.

We are concerned with all the steps regardless of who has investigation responsibility. By breaking the investigative process into these 12 steps we can make the best use of available resources. We can also see that the success of a complete mishap investigation depends largely on top and middle management action. The 12 steps include evaluating the investigative work of subordinates, making valid observations, and taking the proper corrective action, most of which are outside the realm of the investigator.

The essential point of all is: *The thoroughness and effectiveness of any mishap investigation is far more dependent on management actions than on an investigator gathering factual data and arriving at conclusions.* While the manager needs the efforts of the investigator, these efforts are wasted if management is not a part of the total process, from advance preparation to follow-through.

In practice, one finds that the 12 steps are not that clear-cut and readily separated. When a mishap occurs, the steps run together, overlap, and are carried out simultaneously. Understanding the need for preparation and investigation comes first—before the fact. After the mishap occurs, the fact-gathering process starts. At the same time, the reporting, and part of the analyzing, are already under way. Indeed, sometimes four or more steps may be under way at once. This overlapping process need not dismay the manager. If he knows what each step involves, he can still be certain that each one is properly carried out, wherever it falls in the sequence. To this end, let us briefly review the 12 steps in turn.

Step 1. Understand the Need

To be successful at anything we should understand why we need to do it. What exactly is the problem or need? Do we want to prevent future accidents? Are we merely interested in complying with reporting requirements? Will the investigation support us in the event of a liability suit? Are we seeking a scapegoat? Do we want to pinpoint managerial inefficiencies that might be contributing to mishaps and losses? The nature of our findings will be determined largely by what we decide the need is.

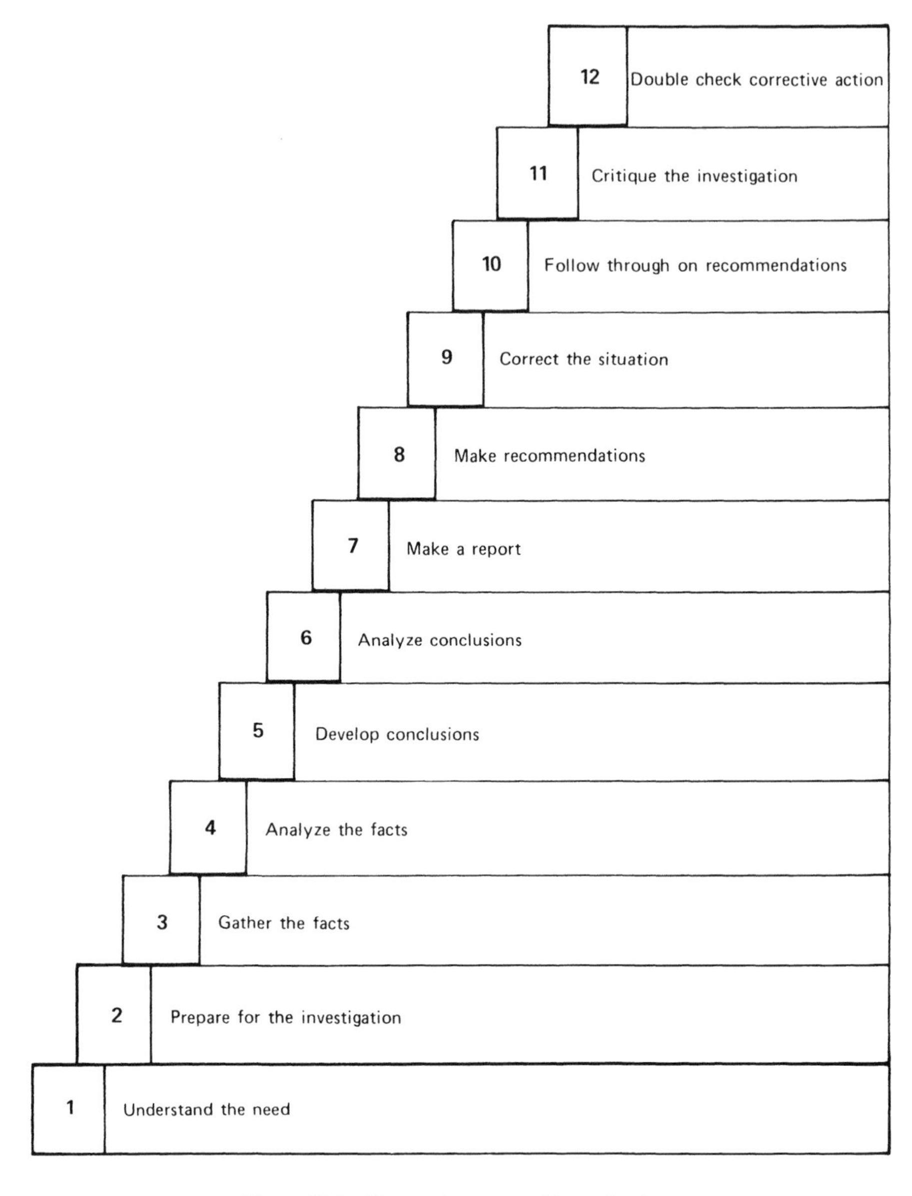

Figure 17.1 The twelve steps of investigation.

It is natural that different findings will emerge for different purposes. For example, if we are investigating the faulty performance of one of our products, we may pay closer attention to what has taken place in engineering or quality control than if we were investigating what seems to be a case of simple operator error. Investigating a construction mishap to meet OSHA requirements would focus our attention on a different area than investigating to find out where management failed.

Perhaps the most important reason for investigation is that mishaps send a message that something is wrong with the way our system operates. Finding the causes and fixing the defects can strengthen the operation, as well as eliminating future mishaps.

Step 2. Prepare for the Investigation

The best approach to handling mishaps when they do occur is to be prepared for them. After the unexpected happens, it is too late for preparation. We need a plan (sometimes called a premishap plan or preplan). This can range from one page to an enormously complex 60 pages. The plan should never be so complex that it hinders prompt action—the briefer the better. Chapter 2 deals in some detail with premishap planning and does not need elaboration here. However, some parts of the plan have special managerial aspects that should be discussed.

A critical part of the plan is knowing that the people who respond to the call for aid at the mishap site may meet hazards. Too many persons are injured at mishap scenes because there are no plans to counter those hazards. It is as much a managerial task to counter those hazards beforehand as it is the task of the on-site investigator.

Having someone definitely in charge at the mishap site is a matter of prior actions by management and staff. It should not be a matter of flipping a coin at the site; there should be a protocol in place that serves when the mishap occurs. The problem is often one of overlapping jurisdictions, since people might operate out of their normal territory or situation. The investigative effort crosses all lines, so plans need to be made for this situation, clearly spelling out who is to take charge.

The preplan should include arrangements for funding or support services such as lab work, transportation, witnesses, photography, chemical analysis, etc. These are obviously things that need to be arranged for before a mishap occurs. The plan also provides procedures for reporting mishaps to corporate headquarters, to family members, and to the news media.

Step 3. Gather the Facts

While steps 1 and 2 are completed well before the actual mishap, step 3 begins after a mishap occurs. The gathering of facts has been fully covered earlier in this book. It is a simple matter if a systematic, clear procedure has been established, detailing what kinds of facts should be collected and how they should be compiled. It is also complex in that the simple procedures may have to be repeated over and over for each new relevant fact or phase of investigation. If this phase is well organized, even the most involved investigations are far easier.

Step 4. Analyze the Facts

The analysis of facts is an ongoing process that begins when we gather the first facts and start mentally weighing them. This mental analysis merges with new information as it is received and may suggest new questions and offer new directions for more fact gathering.

Organization and direction are given this step if everything is well documented and all actions are made a matter of record. While we seldom expect higher management to carry out the analysis, these documents and records can enable the manager to see how thorough the fact-finding was and in turn assess the overall quality of the investigation.

Step 5. Develop Conclusions

As we gather and analyze facts, we draw conclusions about what happened and why. These should be formally organized with the relevant facts on which they were based, to provide a verifiable record. This formal, systematic presentation allows us to see gaps in our knowledge or reasoning. It may point to more areas where more facts and analysis are needed. It will often direct the investigator back to either the fact-finding or analysis, to repeat steps 3, 4, and 5 until adequate and reasonable conclusions are formed. Chapter 12, Multilinear Events Sequencing, is particularly relevant to this process.

Step 6. Analyze Conclusions

This step refers to both our tentative and partial conclusions while we are still gathering facts and our final conclusions after all facts are known, weighed, and analyzed. As suggested above, this is a continuous step that never really stops and may be repeated several times during an investigation. Just as we took time to formally develop tentative conclusions, we should now examine and analyze them from a different standpoint. Eventually, our tentative conclusions will be either made firm or discarded.

Step 7. Make a Report

The mishap report process is well covered in Chapter 16. It brings all our material together—facts, analysis, and conclusions. These facts have been reviewed and analyzed, and conclusions have been advanced, reviewed, and revised. Now all the information is formalized in a report. For some, this is the most difficult and least palatable part of the investigation.

Step 8. Make Appropriate Recommendations

Recommendations are probably the most important part of the report. The finest report fails if it merely states facts and draws conclusions. Corrective actions are needed, and the report should identify them. It should present a specific recommendation for each action needed. That is, each recommendation should cover just one item, spelling out precisely what should be done to correct the situation. Thus the report may list several recommendations, each one stating a specific required action. This allows management to assign specific corrective actions to appropriate people. From this list of recommendations, the manager can say to someone, "Here, this is your responsibility. This is what you need to do. Do it." In this way one person can be held accountable for the correction of each specific causal factor.

Step 9. Correct the Situation

A good investigation should result in corrective action against the causal factors that allowed the mishap to occur. Just placing the recommendations in the report is not enough. The keyword is action. Someone must demand action to keep similar events from happening in the future. A record should be made of who was assigned to perform these corrective actions, for later follow-up.

Step 10. Follow Through on Recommendations

Management cannot rest after recommendations for corrective action have been made, or even after responsibility for taking those actions has been assigned. Someone has to check the corrective action to make sure it has been taken and that it meets the recommendations. Has it been done properly and completely? Have all recommendations been acted on? Were undesired changes made in the process?

Follow-through should include data collection, so that all mishaps can be filed, analyzed, and interpreted for later management decision making. Points to check would include causal and cost information, retrieval of information for data entry, and so on.

Step 11. Critique the Investigation

Many oversights and omissions may occur during an investigation, but the press of getting on with the job often precludes taking action at that time. We may notice flaws in our advance planning, omissions in gathering facts or analyzing them. There may be a variety of goof-ups, particularly in emergency situations. While these problems are still fresh in everyone's minds, before the next mishap occurs, we should critique the investigative process itself, taking action to keep the slip-ups from occurring again. This may be the task of the safety manager, but implementing it will require full management support.

Critique the little things that caused the big problems. For example, the wrong telephone number was posted for the fire department, too many people were allowed to remain around the mishap scene, the supervisor didn't know he should call a member of the safety committee to begin the investigation, the camera used for on-site photos had been borrowed by the personnel department, there was no steno support for typing the report, and so on. These are the types of weaknesses that a critique will bring out. Parties to the investigation should be brought together for this step, since the safety manager will not know about all the problems that were encountered.

Step 12. Double-Check the Corrective Action

Few things get done the way they are supposed to get done. A double check ensures that every action was taken and that recommended changes are kept in force. A one-time performance is not good enough if supervisor and workers revert to the old systems out of habit. The same is true of the problems and corrective action suggested as a result of the critique step. The double check makes certain that everything possible has been done to make future investigations successful and profitable.

MANAGEMENT OVERVIEW

A large-scale mishap involving a fatality, extreme costs, or high public interest will warrant the greatest investigative investment. In addition the corrective actions should be carefully tracked. We cannot always afford the resource expenditure necessary for a large-scale investigation. We can, however, make any investigation as effective as possible by following the 12 steps listed earlier. Our resource expenditures can and should be tailored to the size, nature, and type of investigation, but the 12 steps can always be used.

THE COMPLETE MISHAP INQUIRY SYSTEM

A complete approach to mishap investigation can also be shown as taking place in three phases. This complements the 12 steps of investigation just discussed and everything that has gone before in the book. This approach is based on the idea that the most important things about investigating mishaps are what is done before the event ever occurs and what is done after all the facts are developed. They also happen to be the things most often overlooked in mishap investigation.

Most investigations start with a mishap when it becomes necessary to make an investigation. This step, shown phase 2, the information management phase, in Figure 17.2, is normally considered the most important part of the investigation. The approach used here says that phases 1 and 3 are just as important, if not more so.

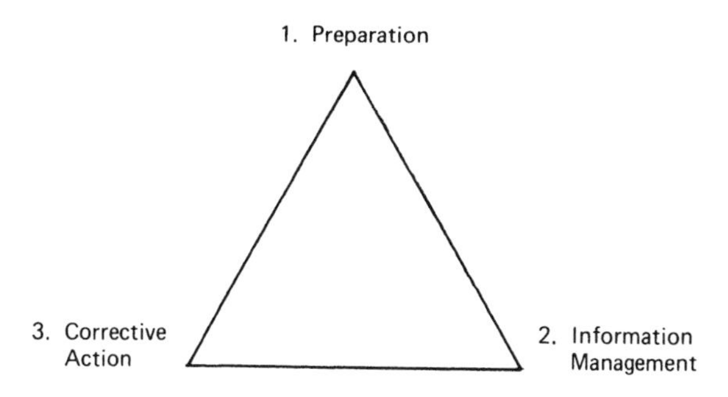

Figure 17.2 The investigation triad.

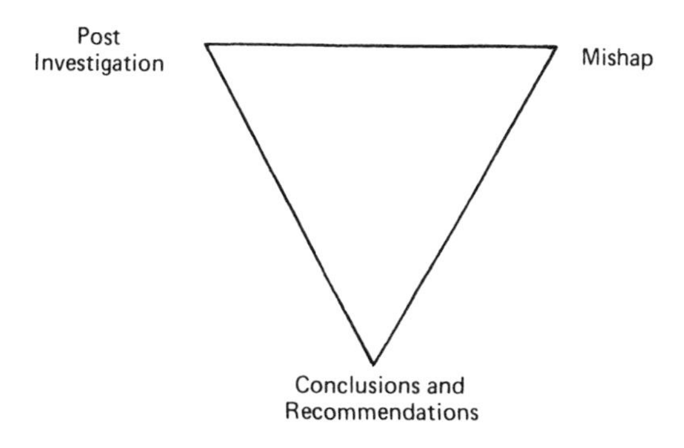

Figure 17.3 Elements of investigation.

Preparation Phase

The preparation phase, completed long before a mishap occurs or an investigation is necessary, is often called preplanning or the mishap or accident plan.

Information Management Phase

The information management phase is usually called the "mishap investigation," but "information management" more accurately describes what goes on during this part of the investigation. Known also as "fact gathering," it must be seen as only a *part* of a complete mishap investigation.

Corrective Action Phase

All information connected with the mishap can be gathered in a highly professional matter by investigative experts. However, if nothing positive happens with that information it is all a waste of time. Hence the most important phase is action, or "completed action," as it is seen here. This action phase is the basis for corrective action.

Many good investigators concede the importance of preparation and completed action but stress that the fact-gathering phase is what makes investigation possible, makes it a success or failure. Without it, an investigation is nothing. However, the proper preparation before a mishap and the proper action once the facts have been gathered assure the best possible investigation and the best use of resources under the circumstances.

The triad of Figure 17.2 does not lessen the importance of fact gathering in mishap inquiries. It emphasizes *preparation* for fact gathering and taking the proper *action* once the facts are known. Our goal is to get the best possible investigative effort and action under the circumstances, regardless of the available resources. To more accurately describe a complete investigation, it is necessary to add some details. This is done with the three additional elements shown in Figure 17.3.

When the second triangle is superimposed on the first, we have a star that looks like Figure 17.4. The interacting lines of the star show the interface of the various stages as well as a definite sequence built around the three phases of preparation, information management, and action. At the top is the preparation for investigation. A mishap then triggers an information management stage, leading to conclusions and recommendations. The action step does the proper thing with findings and conclusions. Finally, to tie everything together, there is another step called the post-investigation stage. This process may also be viewed as in Table 17.1.

Preparation

These are the things to know and be prepared for before a mishap ever happens.

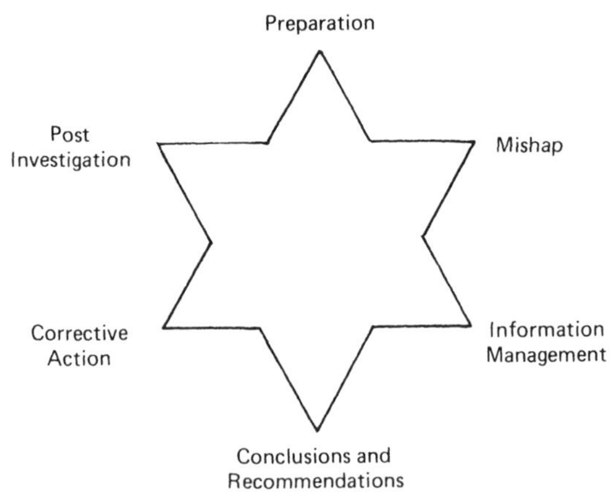

Figure 17.4 The investigation star.

Table 17.1 Three Major Phases of Complete Investigation

Preparation	Information Management	Corrective Action
Objectives	Develop	Forms
Effort level to:	Organize	Correction
Record	Integrate	Follow-through
Report	Conclusions	Evaluation
Investigate	Recommendations	Feedback
	Report	

1 Objectives: Know the objective of the investigation.
2 Effect level: Decide the level of resources to be used.
   (*a*) Record: Know how information will be gathered and recorded.
   (*b*) Report: Know what type and level of reporting will be used.
   (*c*) Investigate: Know how investigations will be carried out.

Information Management

This is where the facts are gathered, developed, and evaluated for their role in mishap causation.

1 Develop information: Gather facts and consolidate them.
2 Organize information: Develop causal relationships.
3 Integrate information: Coordinate and look for knowledge gaps.
4 Develop conclusions based on integrated information.
5 Develop recommendations to satisfy investigation objectives.
6 Make report: Use a viable format leading to action.

Action

Without action after all factual information is developed, the resources used in investigation are wasted. The purpose of the investigation—prevention, litigation, find costs, or whatever—is not accomplished. The investigation effort is in vain if action is not taken on the mishap report.

1 Forms: Proper distribution to parties needing to take action.
2 Corrections: Activity by single, specific, responsible party.
3 Follow-through: Check assigned corrective activity.
4 Critique: Of the complete investigation and correct discrepancies.
5 Feedback: On findings, discrepancies, and corrective actions.

SUMMARY

A complete investigation making the best possible use of available resources requires preparation for a possible mishap investigation before a mishap ever happens. After the mishap is investigated and factual, causal findings and conclusions are reached, a report is made. This is not the end of the matter, merely preparation for the most important part of the investigation: taking action to prevent similar and related mishaps in the future. Without the action step, an investigation is incomplete because it lacks results. Thus, a complete investigation takes us from preparation for a mishap through completed action to correct all discrepancies and omissions, and finally to critiquing the investigation process itself to ensure that it goes as expected the next time investigation is needed.

QUESTIONS FOR REVIEW AND DISCUSSION

1. Which of the 12 steps of investigation is the hardest? Why?
2. Of the 12 steps of investigation, which ones deal with investigation as most people or agencies describe it? Justify your answer.
3. Why should we be concerned about a complete mishap investigation package that starts with planning long before a mishap happens and continues well past the time the mishap report is finally turned in? Let's be practical. Shouldn't we just get the facts around an event and turn in our report?
4. Name the three major phases of a complete investigation. Describe in a short paragraph, in your own words, what each phase means.

QUESTIONS FOR FURTHER STUDY

1. The National Institute for Occupational Safety and Health has endorsed an accident investigation technique called CAIRS. Find out what CAIRS stands for and briefly describe it.
2. Compare CAIRS with the ideas set out in this chapter, and list the shortcomings of this chapter's approach in comparison.
3. In Chapter 12, we referred to the S-T-E-P approach to investigation. Through library or personal research, determine if the approach suggested by this chapter is compatible with S-T-E-P.

REFERENCE

Safety Management Planning Manual, Ted Ferry, Ed., The Merritt Company, Santa Monica, CA, Chapter 11. Extracts of material from this source have been used here with the permission of the Merritt Company.

CHAPTER EIGHTEEN

Legal Aspects of Investigation*

The purpose of this chapter is to acquaint the investigator with various problems, limitations, and requirements that arise in a typical mishap investigation. It cannot be a comprehensive in-depth dissertation on the law of mishap investigation, but its purpose will be served if it triggers the reader's curiosity about those legal topics that directly concern him or her.

MISHAP PREVENTION VERSUS LITIGATION CONSIDERATIONS

In mishap investigation, the investigator is faced with the dilemma of competing goals and interests of safety, mishap prevention, and possible future litigation. Foremost in the mind of the mishap investigator is public safety and the prevention of future mishaps. A statement in a mishap report that a product is defective and dangerous and should be immediately recalled from the field and removed from the hands of all users may be used as an admission of liability in the present mishap. It may be used as evidence to convince a jury that the company should be liable to an injured plaintiff.

Our legal system is adversarial, based on the philosophy that if opposing sides fully develop and argue their respective positions, the truth will emerge. The mishap investigator is typically not trained to function in such a system. The investigator sees his responsibility as the unbiased search for truth, for the causes of the mishap and the most effective way to avoid future occurrences.

The two are not as unreconcilable as may appear on first blush. As mentioned earlier, mishap investigation must be directed to finding facts, not faults. The goal of the mishap investigator is to determine what is known in the law as "technical causation." Technical causation is the relationship between manufacturing flaws, management deficiencies, human error, design deficiencies, inadequate warnings, and the mishap. In other words, did the deficiencies cause the failure or malfunction of the product or induce the human to cause the mishap?

* This revised chapter was originally written by Lewis Bass, Attorney-at-Law, Mountain View, California.

ROLE OF THE MISHAP INVESTIGATOR

The investigator evaluates the evidence and performs tests and analyses with the utmost scientific integrity to answer this technical causation question. He or she should not venture an answer related to the ultimate question of legal responsibility. It is the role of the judge to determine what limitations should be placed on a defendant's responsibility for mishaps that are technically caused by his products or employees. For example, a product may have technically caused a mishap but no liability be placed on the manufacturer because the injured party is shown to have misused the product. Finally, it is the role of the judge or jury, the trier of fact, to apply the principles of law to the facts and decide questions of responsibility.

Fact Gathering

The investigator is usually the only person at the scene concerned with reconstructing the mishap and preventing future mishaps. Others at the scene are concerned with the welfare of the injured victims and with public safety. They may inadvertently destroy evidence and facts needed to find the causes of the mishap.

Evidence Preservation

It is the investigator's responsibility to properly preserve, label, and document the location and identity of evidence. It is critical to document the scene as close to the time of the mishap as possible. Evidence that has been exposed to the elements for a lengthy period of time may no longer provide the facts needed to determine the causes of the mishap. For example, skid marks may disappear. The scene of a mishap may be also cleared or the area rebuilt and otherwise changed.

Access to Evidence

If the evidence is on the company's own property, there is, of course, no problem of access. However, if the evidence is on the property of the adversarial party or is at a location unavailable to the investigator, the investigator should be aware of certain legal limitations on his actions. You cannot enter others' property without their permission. Doing so will expose you to a charge of unlawful trespass. If permission is not granted, do not attempt to investigate any further without securing legal clearance.

There is a legal procedure whereby an attorney may request that the investigator be allowed to inspect the scene or the product. "Inspection" generally includes taking photographs and measurements and making non-destructive tests. There is also a process whereby the party having possession of the evidence is required by the court to preserve and protect it.

Testing

To carry out a full investigation of the mishap causes, the investigator may need to do destructive tests on the product or evidence. Do not do so without first getting the advice of counsel. Both sides have the right to inspect the evidence in its original condition. To unilaterally and destructively test the evidence would impinge on that right. Such tests may only be performed with the permission of both sides or by direction of the court. Representatives of each side are allowed to be present at the testing and to obtain the facts relating to the procedure and the results of the tests. Using laboratories and testing services can be a time-consuming and expensive process. But not doing it may also be very expensive.

When tests are necessary they should be performed by an accredited and independent testing laboratory. If the tests use company facilities and personnel, an independent and unbiased observer should be present. The credibility and validity of the tests will be an important issue in any legal proceeding. Proper care must be taken to document the test procedures. Instruments, apparatus, and equipment used in performing the tests must be properly calibrated. These calibrations must be traceable to the National Bureau of Standards and properly documented. The source and purity of chemicals used in performing the tests should also be documented, as should the specifications of any other equipment used. The importance of these rules cannot be overemphasized.

Photos, Sketches, and Measurements

The scene will change shortly after a mishap. It is imperative that the investigator quickly record all pertinent facts. This includes taking photographs of the scene and of all pieces of evidence. The photographs must be properly labeled as to the object photographed, the location, the time taken, the person who took the photograph, and a notation of the scale. It is not necessary for the person who actually took the photographs to appear in court. A person familiar with the scene and the evidence can testify on the identity and description of the contents of the photographs.

Sketches should be made of the scene showing the location of all victims and evidence. If the sketch is not to scale, it should be so noted. Sketches made by others can provide useful information, but they should never be relied on without independent investigation and measurements.

Chain of Custody

There are often many people at the scene, such as police and fire department personnel, who have access to the evidence and who pick up information or cause evidence to be stored in some form. It is vital to know what this material is, where it is stored, who is in charge, and what process is necessary to view the evidence or obtain possession. To introduce evidence, the chain of custody of the evidence from the time it was retrieved to the date of the trial must be given. Documenting the chain of custody involves identifying the location of the evidence when it was retrieved and all persons who came in contact with it until the time it was presented to the court. The court must be assured that the evidence is in its original condition or that all alterations in its original condition can be explained.

Interaction with Witnesses

Witness interviews are a potential problem area for the investigator. Always be honest with the witness about your identity and status. Do not misrepresent your employment or your authority. In the trying time shortly after a mishap, the investigator must be careful not to cause the witness additional emotional harm or distress and to respect the witness's right to privacy. Do not harass the witness by contacting her or him at inappropriate hours or interfering with normal activities. Interview the witness in comfortable, familiar surroundings. The witness's home is better than his place of work. If the witness does not wish to be interviewed, the investigator must not press the issue. There is a legal process known as the "deposition" whereby the witness can be interviewed by counsel under oath and his story taken down by a court reporter. If the witness has given a statement to the opposing party, there is a process whereby the attorney can determine its substance or obtain a copy. The investigator should know that he is part of the discovery team. When you need help to uncover facts, you should seek advice from other members of the team before proceeding in what may be an unlawful manner. Otherwise you could subject yourself to criminal or civil penalties.

Licensing

Because of the potential abuses of unethical mishap investigators, some states have imposed licensing requirements. In California, for example, an independent investigator hired to investigate the facts surrounding a mishap must comply with the applicable business and professional codes.

Before an independent investigator practices in any state, he should be aware of local licensing requirements. There are various exceptions to these requirements. An investigator who is employed exclusively by one employer or who is an attorney-at-law is not required to be licensed in California. There are civil and criminal penalties for investigating without a license where one is required, and the investigator may be kept from testifying on his findings.

Witnesses' Statements

The investigator must decide whether to record the witness's statement on tape or obtain a written statement. Most witnesses are ill at ease with recording devices. The investigator must decide in each situation which approach will be more effective.

Written Statements

When preparing the statements, use short, concise, simple language that is appropriate to the witness. Avoid technical or legal jargon unless the witness would naturally use such language. The validity of many statements is questioned when it is obvious they are not in the witness's own words. Be certain that the witness reads the statement and signs or initials the bottom of each page.

Recorded Statements

Before recording a witness's statement, the investigator must have the witness's permission. Statements recorded without permission are an illegal invasion of privacy. The investigator may be subject to sanctions, and the evidence will be inadmissible. Begin the secondary by stating your name, the time and location of the interview, and that you have asked for and received permission of the witness to record the interview. A recorded statement is evidence and must be preserved, even if you only made it to help you write your report.

The purpose of the interview and statements is to cement the witness's recollection at an early point in time. Memories fade, and the witnesses' future recollections may be confused by newspaper accounts and discussions with other witnesses and fellow workers. It is imperative, therefore, that the witness be interviewed quickly after the mishap. The statement also preserves the witness's testimony. The witness may be infirm or aged or may move residences or otherwise disappear and be unavailable when needed at a later time. To preserve such evidence, the investigator must inform the attorney of these conditions so that the attorney may quickly take deposition of the witness and preserve the testimony. If the witness is unavailable at a later time, the deposition can be used instead.

Another purpose of the witness's statement is to point out any inconsistencies between what the witness said to the investigator and later testimony at the deposition or trial. In order for the answer to be of value, the questions should be simple and unambiguous so that the meaning of the witness's response is clear.

DISCOVERABILITY AND PRIVILEGE

Discoverability of the Investigation Report

The discoverability of the investigation report depends on the investigator's status at the time of the investigation and at the time of trial. The entire report of a company-employed investigator is discoverable by the opposing attorney. This includes the facts and opinions, conclusions, and recommendations of the investigator.

It is a basic rule that discovery of relevant material is limited by the rule that "privileged" matters are not discoverable through any discovery procedure. Information that is the result of the attorney's work product is not discoverable unless the court determines that denial of discovery will unfairly prejudice the preparation of the moving party's claim or defense, or will result in injustice. A writing that reflects the attorney's impressions, conclusions, opinions, legal research, or theories is not discoverable under any circumstances.

Privilege may be waived by voluntary disclosure of the privileged information, by failure to assert the claim of privilege, by tendering certain issues, or by conflict between discovery and trail that is inconsistent with such claim.

If the attorney asks the investigator to pursue the investigation in support of, or in preparation for, the case, then different rules apply. The attorney is protected by what are known as "attorney–client" and "work product" privileges. These are limitations of the otherwise liberal rules of discovery. They are based on the concept that each attorney must prepare his own case and cannot sit idly by and have the other attorney prepare the case for him. Similarly, information provided by the client to the attorney is privileged from discovery in the interest of full and truthful giving of facts to an attorney.

A client, whether or not a party to the action, may refuse to disclose and may prevent another from disclosing a *confidential* communication between himself and his attorney. The client may be an individual, partnership, corporation, public entity, group, etc. A confidential relationship must be intended in order for a communication to be privileged.

These privileges extend to the agents of the attorney who assist in preparing the case. Thus, when the attorney calls on an accountant, real estate appraiser, or investigator for assistance in preparing a case, the opinions, conclusions, and recommendations of these agents are given the same privilege. Factual information obtained as part of the investigation is not privileged. Photographs, for example, are discoverable by the other side.

Consultant or Potential Expert?

The conclusions and opinions contained in the report of an investigator or expert retained by an attorney are privileged until the investigator or expert becomes a potential witness. At that time, the work product privilege is waived. When the attorney declares that the investigator will be a witness at trial, the investigator is then available for deposition, and his report is available to the other side.

When Is the Investigator Deemed an Expert?

An expert is a person with specialized knowledge in an area beyond the competence of a lay jury. For example, more jurors cannot interpret skid marks to decide the speed of an auto before the collision. A police officer, mishap investigator, or reconstruction engineer who is specially trained and skilled in making such determinations will be asked to give an opinion or conclusion along with the basis for it. The expert is an individual who because of training and experience has special knowledge of a specific field. The law recognizes the expert as special and distinct from the usual witness or observer.

The expert need not have any special academic qualifications if by experience and training he shows the ability to give a knowledgeable opinion. The background of the expert does have an effect on the weight that the jury places on his opinion. If the accident investigator has academic training, many years of experience, is the author of a textbook on the subject matter, and is active in professional societies, the jury may place greater weight on his opinion than on that of an expert with lesser qualifications. Similarly, the validity of the conclusions and recommendations of the mishap investigator will depend on his qualifications. The investigator will be asked about his professional experience and his particular expertise with the product, process, or facility under investigation.

The expert will be questioned about the facts he considered and the documents he relied on in forming his opinion. He will be asked questions about the tests he made, analyses he performed, and his reliance on the opinions of any other experts or treatises.

One type of question that an expert can answer and that cannot be asked of an ordinary witness is the "hypothetical" question. The expert is asked to assume a hypothetical set of facts. These facts are based on facts the jury could find from the evidence given or to be presented in the case. For example, he may be asked to assume hypothetically that a pressure relief valve malfunctioned and to give his opinion on whether such a malfunction could have caused the accident in question.

Using Experts

When should you call on an expert? The decision usually comes when it is obvious that some technical problem is beyond the scope of your experience or you require additional support to sustain a decision regarding a claim. If something is amiss that may be a borderline technical item, it is better not to take a chance. Call an expert. However, you should not use an expert if the potential gains do not justify the expense. In that case, is better to seek a settlement.

Acting as an Expert

There are increasing opportunities for professionals to serve as expert witnesses. To be effective you must first analyze the nature and function of the forum (court, hearing, OSHA, etc.) in which you are testifying and your role in it. The attorney with whom you will be working can help clarify these matters. By virtue of your specialized knowledge, as an expert witness you will have the opportunity to express opinions and conclusions that are normally beyond the knowledge or experience of the average witness. The expert need not testify from direct observation of certain events. Many times the outcome of a controversy will be decided by a contest between opposing expert witnesses who are asked to comment on hypothetical situations.

To be an effective expert witness, you should follow the five C's:

1 **Credibility.** Establish credibility by making a good first impression, displaying modesty, good humor, and patience. Do not be condescending or overly casual.
2 **Correctness.** Do your technical homework, and be current and conversant in your information. Prepare notes to assist your memory but use them as little as possible when testifying.
3 **Clarity.** Be clear in the content and delivery of your material, speak slowly and with confidence, and avoid technical jargon. Use words you think the judge and jury will understand.
4 **Conciseness.** Say only what is necessary to make your point, and then stop. Avoid repeating yourself unless asked to. Do not volunteer more information than is asked. Do not venture beyond the bounds of your qualifications. Avoid getting into a verbal exchange with the opposing attorney.
5 **Candor.** Be natural, sincere, and honest; show respect for your opponents without conceding your position. Remember: You win when the jury elects to accept your testimony over your opponent's.

TESTIMONY OF THE MISHAP INVESTIGATOR OR EXPERT

The Deposition

Under the rules of practice, each side in a lawsuit has the right to take the deposition of the opposing party and potential witnesses. A deposition is the oral testimony of a witness taken under oath before trial and recorded by a court reporter. The purpose of the deposition is to determine all facts that the witness may have in his possession that will assist the attorney in the preparation and trial of the lawsuit.

At the deposition or at the trial, the listeners' impressions of the witness may be important considerations in the weight given to the testimony. The investigator must present the impression of one engaged in the unbiased scientific pursuit of truth. As an expert witness, you must, above all else, honestly and fairly tell the truth. Do not be an advocate. If you do not understand a question, request that it be explained. Never guess at the answer to a question you do not fully understand. Answer directly, giving concise answers to all the questions. Wait until the question is completed before you answer it. Stick to the facts, and testify only on what you personally know. Most of all, if you do not know the answer to a question, admit it. Some witnesses think they should have an answer to every question asked. You cannot have all the facts, and you are not answering accurately if you attempt to testify to facts with which you are not acquainted.

Trial

At trial there are two phases of testimony: direct examination and cross-examination. During the direct examination, the investigator presents his story to the best of his ability, relating what he saw, did, and concluded. The opposing counsel then has the right to question the witness on his testimony. Opposing counsel will attempt to show inaccuracies in the witness's facts or deficiencies in his conclusions. Above all, the investigator should neither lose his temper nor be afraid of the attorneys. There may be a genuine difference of opinion in the scientific community on the conclusions that may be drawn from the evidence. If there were no differences of opinion about the relative responsibilities of the parties and the cause of the accident, there would not be a lawsuit. It is imperative that the witness be honest and straightforward in his testimony. The opposing counsel is not being personal in his cross-examination but is trying to present his side to the best of his ability.

MISHAP REPORTING

Reporting to Company Management

To have any hope of protecting prelitigation inspection or test reports, it will be necessary to have the services of an attorney. If, as a result of the mishap investigation, a safety problem is exposed in a product or facility, company management must be made aware of the need for corrective action or other remedial measures. If a future mishap occurs and no action was taken, punitive damages (which are generally not recoverable) may be awarded to the injured party as a punishment for willfully and wantonly placing the public in danger of a known hazard. Immediately after a mishap:

1 Secure the mishap scene.
2 Call the company attorney for directions so any witness statements and photographs taken will, hopefully, be protected by the attorney work conduct privilege.
3 Collect physical evidence as appropriate.
4 Transmit witness statements and photographs to the attorney.

These steps do not guarantee privilege for the reports and statement, but it is better than doing nothing to prevent their disclosure.

There are occasions when the findings of the expert may not be favorable to your side. This is usually reported to the attorney verbally, with the question of what to do now. This information may be vital in the early settlement of a claim, but you may not want it discoverable at some later date. The expert's notes made while preparing a report can be obtained and used by the opposition if the case goes to court. Therefore your attorney may ask for "no report."

If there is the slighest suspicion of criminal activity discovered by the expert, he should inform his attorney and then call the proper law enforcement agency.

Regulatory Agency Requirements

The investigator should be aware that there are mishap reporting requirements under various federal and state acts.

Consumer Product Safety Act

Section 15(B) of the Consumer Product Safety Act requires that every manufacturer of a consumer product distributed in commerce, and every distributor and retailer of such, who obtains information that reasonably supports conclusions that such product:

1 Fails to comply with an applicable product safety rule or
2 Contains a defect that would create a substantial product hazard (a product defect that creates a substantial risk of injury to the public)

shall immediately inform the Commission of such failure to comply or of such defect.

If, during your investigation of a consumer product, you become aware that a product creates a substantial risk of injury to the public, you must immediately report these facts to the Consumer Products Safety Commission. There are civil and criminal penalties for failure to conform to these requirements.

Medical Devices Act

Under the Medical Devices Act (Section 820.162) as amended in 1976, there are detailed requirements for conduct of an investigation of the failure of a medical device. The act requires that failure of a medical device or any of its components to meet performance specifications shall be investigated. A written record shall be made, including conclusions and follow-up.

Occupational Safety and Health Act

The Occupational Safety and Health Act likewise imposes recordkeeping and mishap reporting requirements on employers whose employees sustain an injury during the course of their employment. These requirements are detailed in Chapter 16.

SPECIAL CONSIDERATIONS OF THE MILITARY AIRCRAFT MISHAP INVESTIGATION

In the military, there are two separate mishap investigations, the safety investigation and the collateral investigation. The purpose of the safety investigation is to draw conclusions and make recommendations for preventing future accidents. These conclusions, opinions, and recommendations are privileged for disclosure in any future litigation. The privilege of nondisclosure granted in safety investigation reflects the public policy of determining all the facts relating to a mishap in the interest of public safety. The pilot, mechanic, or air traffic controller who tells the safety investigator that he made an error, or the manufacturer who acknowledges that he supplied a defective part to an aircraft, does so fully and without fear of reprisal or criminal or civil liability as far as the safety investigation is concerned.

The collateral investigation has as its purpose the preservation of evidence and testimony and covers all mishap investigations other than for safety (prevention) purposes. The collateral investigation report may be used in subsequent litigation. The manufacturer of a consumer product who reports a potential "substantial product hazard" to the Consumer Product Safety Commission has no assurance that under the Freedom of Information Act this information will not be supplied to the plaintiff's attorney.

SUMMARY

In conclusion, the mishap investigator must be aware of the limitations placed on his ability to gather facts and the importance and impact of his opinions, conclusions, and recommendations on future litigation. He must know the witness's right to privacy and the penalties for misrepresentation or trespass during the investigation. The investigator must be aware of the reporting requirements of government agencies. He must also make the company management aware of any critical safety deficiencies in their product or facilities.

QUESTIONS FOR REVIEW AND DISCUSSION

1. Explain how the investigator is faced with competing goals regarding mishap prevention and litigation.
2. What is meant when we say that our legal system is an adversary system?
3. Who usually has the responsibility to preserve all mishap evidence?
4. Is there a legal means to allow the investigator to inspect mishap evidence in someone else's possession? What is it?
5. What is meant by the "chain of custody?"
6. Explain discoverability.
7. When is the investigator deemed an expert? Explain.
8. Identify the two phases of testimony and summarize the investigator's role in each.
9. Comment on the requirement of the investigator to report to the Consumer Product Safety Commission when he is aware that a product creates a substantial risk of injury to the public.

QUESTIONS FOR FURTHER STUDY

1. Can you get a witness for the opposition to give your side a statement? How?
2. Research the licensing procedures for investigators in your state and explain them.
3. Research the Medical Devices Act of 1973 and its 1976 amendment in enough depth to outline its major provisions.

GENERAL REFERENCES

1 Bass, Lewis, *Products Liability: Design and Manufacturing Defects*, Shepards/McGraw-Hill, Colorado Springs, CO, 1986.

2 Cornwell, Dale, "Privilege," undated lecture notes from the University of Southern California, Los Angeles, CA, 1986.

3 Golec, Anthony M., *Techniques of Legal Investigation*, Charles C. Thomas, Springfield, IL, 1976.

4 *The Adjusters Guide to Forensic Engineering*, Garrett Forensic Engineers, Los Angeles, 1986.

5 McGrew, D. R., *Traffic Accident Investigation and Physical Evidence*, Charles C. Thomas, Springfield, IL, 1976.

6 "You Can Be An Effective Expert Witness," *Professional Safety*, February 1986, pp. 28–30.

7 Weinstein, A. S., A. D. Twerski, H. R. Piehler, and W. A. Donaher, *Products Liability and the Reasonably Safe Products*, Wiley, New York, 1978.

CHAPTER NINETEEN

What Is Ahead for Investigation?

The future for investigation can only be bright and encouraging as we approach the 21st century. The coming decade is seen as one that will involve tough-minded approaches to business and industry directed by well-schooled executives and managers of exceptional ability. We continue to be in a period of tight competition involving new technologies and innovative management. The challenges to the safety professions have proved to be arenas in which to thrive. A new breed of investigators now coming on the scene will benefit from the new technologies and new management. Investigation will be more efficient and yield greatly improved results in the more demanding environment.

PRACTICAL PROBLEMS WITH INVESTIGATION

Despite a bright outlook for better investigation and new investigative techniques, the present mishap investigation system is of immediate concern. Problems with the present arrangements of mishap investigation often interact. In a synergistic sense the total adds up to a dim picture. There are a few exceptions in specific organizations. The following are common problems.

1. Investigations are usually conducted by untrained people.
2. Management does not recognize their stake in investigation and thus has superficial interest.
3. Reporting formats provide data for tabulation instead of correction and prevention.
4. Management feels that there are not enough resources to properly investigate mishaps.
5. The organization in general does not feel supportive of good investigation.
6. The benefits of good investigation are not recognized.
7. Good investigation is too costly.

8 More benefit may derive from review of the investigation of others.
9 No one is practically and readily available to train organizational people in investigation.
10 The government that sets the regulations is only mildly interested in investigation of mishaps.
11 Good investigative techniques are not widely known among those seeking to improve their techniques.
12 Good investigative consultants are too busy making a living to make general contributions of their knowledge.
13 There is no journal of investigation to pass along knowledge, except in a couple of highly specialized areas.
14 Few organizations attempt to incorporate mishap findings into active prevention programs.
15 No one is charged with making specific recommendations for correction based on mishap investigation.

The above list addresses general situations. There are always exceptions and/or methods of correcting them. Nevertheless, these problems are largely valid, and several of them may exist in an organization, either actually or as perceived. They combine to form a dismal picture of mishap investigation. They are all at least a little valid everywhere and are found in nearly all organizations.

This glimpse at present problems provides a springboard to look at what is ahead for investigation.

CONCERN WITH LITTLE MISHAPS

It has been suggested from time to time that investigation techniques for a sophisticated and complex mishap have little in common with what appears to be a minor incident or even a near miss. To the contrary, the "little" event may have all the implications and complexities of a catastrophe. It only appears less important because serious injury or property loss did not result. The same causes and modes of failure are present in both large and small mishaps. Our techniques have focused on the more spectacular, more costly, and more mysterious mishaps that at first glance seem to defy solution and thus call for a "real" investigation.

There are hundreds of commonplace mishaps that in passing seem to have little in common with high-technology multifaceted mishaps. This suggests that those with the greatest loss of resources are more important, but it is only that they are the events that demand immediate attention. In actuality, mishap results are largely fortuitous, but the causes are similar. A relatively simple electrical mishap in a facility dealing in exotic metals may result in a little shock or minor delay. But under the right circumstances it could result in a major loss of resources with significant loss of life.

In either case, the causal factors that made the mishap possible are largely identical. Perhaps the presence of only one more factor would have resulted in the worst possible case. Is there a good reason to concentrate only on those mishaps that result in a major loss of resources? The less significant event may have all the causal factors present in an environment far easier to investigate. Since there are many more of the lesser events, the opportunity to find and fix causal factors with a smaller expenditure of resources should be used to prevent future mishaps that would have more serious consequences.

Consider the situation where a machine is being repaired and an electric spark is generated. In most situations this is all that happens; no one is hurt, and there is little if any damage. However, the destructive potential of the event is great. It should be investigated with the same spirit of inquiry as if a major resource loss had resulted. If the spark had been in a grain elevator or near a carbon-based materials processing operation a major catastrophe could have resulted.

We should consider the wisdom of investing resources only in the prevention of catastrophic events as compared to those with less significance. Since investigation is a prevention tool, the inquiry into minor mishaps may not be considered a good use of resources. If even the minor mishap demands thorough inquiry of an appropriate amount, what then is an appropriate inquiry?

1. If the operating environment is highly hazardous, then prevention efforts (investigation included) deserve a high resource investment to prevent a major loss. High-hazard operations enjoy the same causal factors as low-hazard operations. Only the loss of resources or the likelihood of mishap is different. Since the amount of resource loss is often a matter of chance, even the low-hazard operation deserves careful attention.
2. What may seem to be a small loss potential is often high. For example, common back or eye injuries may be given scant attention because of their simplicity and extended absence from the job is covered by workers' compensation. We simply substitute another worker. The simplicity of these mishaps tends to discourage thorough investigation and corrective action. In reality we are dealing with two of the most expensive injuries in terms of hidden expense. If there is spinal cord damage from the back injury, the cost averages out to more than that of several deaths.
3. Sophisticated investigators and managers know that most mishaps, particularly those common minor events that only cause disruptions, inconvenience, and delays, are symptoms of inefficient management

and operations, and they will look to the thorough investigation of each as good business.

PROBLEMS AND BENEFITS OF TECHNOLOGY

New technologies create new hazards. This trend is now seen in the areas of toxic and radioactive substances, new metals, and electronics. Fortunately the imaginations of those called on for investigative assistance have responded with new approaches to the task. For each new threat we develop new investigative tools and techniques. For example, a test to determine the presence of a dangerous substance is also an investigative device to detect the presence of the substance after a mishap.

Newer families of business equipment will have a favorable impact on investigation. Interdisciplinary investigation will be less of a mystery and far easier to carry out as the new machines bring on tightly coordinated, timely, and comprehensive investigation management. The data developed for all areas of business and industry will provide more resources for solving mishap problems. There is now desktop access to the work of scientists and engineers at a touch of a few keys. Mathematical analysis, modeling, and graphics are becoming readily available tools for the investigator. The same automated systems that provide product and maintenance information are available for investigative research and will be a common input in the future.

Major problems will surface with the new machines and their capability to provide information. It may be the wrong information, or the special aspect needed will not be available for our investigation purposes. Unless there is a reversal of trends there will continue to be inhibition in information exchange through the new machines due to possible legal actions arising from private litigation and government regulation.

THE LANGUAGE OF MANAGEMENT

More than ever our investigations will pinpoint management and human factors as causal aspects of mishaps. Their complexity, a drawback to past use, can be handled by computer interface. Their complex interrelationships will be easier to understand and visualize when a computer can, for example, present them in matrix form. In those matrix organizations the interfaces may be recalled quickly. Imagine yourself trying to trace the design, acquisition, delivery, storage, and use information on a failed bolt that had figured in a mishap. A push of the buttons and we secure the data we need, along with cost–benefit information on using present stocks and on converting to new bolts. Already this can come to your desktop in a matter of seconds.

There are few times when management needs help more than when there has been a significant and perhaps costly mishap. Everyone and everything is putting pressure on the principals of the company. To list a few of the pressures:

1. There has been a disruptive event calling for some type of action. What action should be taken?
2. Every mishap is accompanied by chaos, and order must be restored.
3. Work processes are disturbed and sometimes shut down.
4. Profit margins shrink or vanish.
5. Expensive machinery may be damaged or lost.
6. Dollar losses will exceed insurance payments.
7. Respected members of the organization may be lost.
8. The news media are clamoring for information, and bad press will damage the corporate image.
9. Families of survivors need attention.
10. Public confidence in the organization will be shaken.
11. Stockholders and members of governing boards will demand explanations.
12. Governmental investigations with attendant finger-pointing and time-consuming demands for attention will result. Violations of regulations may surface.
13. Doubt will be cast on similar operations in the organization, suggesting that operational, supervisory, and management changes may be in order.
14. Insurance and workers' compensation costs will probably rise as a direct result of the mishap.
15. On-the-fence customers will jump to other suppliers.
16. Top and middle management will spend long and valuable hours seeing the affair through a successful investigation and corrective action.
17. The morale of the work force will be adversely affected.

These are only a few of the management concerns when a mishap occurs. It is clear that management is concerned. Whoever is doing the investigation, making recommendations, and taking corrective action must do it in management language. If you, the investigator, can speak this language and can translate investigation jargon into answers to the above concerns, your future is assured. Only one person is positioned to bring order out of the items listed above. The immediate person on the spot is the investigator, who through good investigation finds the supervisory and management deficiencies involved and makes suitable, practical recommendations that can be acted on. If he does it in management terms, he becomes a savior on a white horse to a harassed management.

THE NEW LOOK OF INVESTIGATION

The style of an investigation is usually keyed to the operational style of the organization. For example, investigation will be directly influenced by the rapid movement of business and industry into electronic business equipment. Hardly a new field, recent developments and gradual acceptance have assured its place as an investigative tool. Since business and industry will be involved, as a matter of survival the common language of electronic business equipment will become the language of the investigator as well.

The advantages to the investigator will be many. We can expect the electronic processing of information to result in higher quality decisions because of data available to the investigator. There are hundreds of data bases of general and specialized value. They range from legal and scientific abstracts to financial data. They are largely available through retail intermediaries, and their direct access is growing. Thus the programs and data bases needed for mishap investigation will be readily available. Additionally, there are tens of thousands of microfilmed data bases for nearly every subject.

The investigation data bases now cover such things as human factors, machine factors, severity of mishaps, and environmental factors. Because of the requirements of law enforcement officials, the legal profession, medical personnel, and others, the basic source documents are overloaded with data not of direct interest to the investigator. Federal agencies are trying to bring some order out of the situation by standardizing collection modes.

Soon everything we put on paper or in the computer will be kept in computers and data banks. For those of us who cannot recall exactly what we worked on and when, pressing a few keys will bring the information to display or printout.

Compliance with regulations and other requirements will not be guesswork, but readily available on a screen, even integrated into a format. This will allow more efficient processing of investigative data and should end investigation overkill, reduce the administrative overload, and increase the value of the information received. Unfortunately, GIGO (garbage in, garbage out) will still apply, but if reporting requirements are properly managed this may be the big breakthrough we seek without overloading reporting parties.

It is too difficult and time-consuming at present to keep track of resource expenditures during a mishap investigation. The new machines will allow us to quickly, accurately, and easily track our expenditures. This may force concentration in certain areas but will also increase efficiency by guiding our effort through continual feedback of results. Much of what was guesswork on the efficiency of investigations will become a matter of record, thus showing the way to efficient allocation of our resources.

A preoccupation with finding data for inserting into the electronic system may combine unfavorably with a tendency to seek only factual material in investigations. This may lead us to the computer-age syndrome of knowing everything but understanding nothing. Room must be left in the system for interpretation and the mulling over and consideration of conjectures and speculations. Electronic processing must never become the master of the investigation by interfering with the discovery of elusive but vital information and interrelations.

Efficient and Orderly Investigation

An efficient investigation is one that secures essential information for durable corrective action with a minimum use of resources. It calls for investigating what is necessary but no more. It also calls for investigating in depth but not beyond the point where corrective actions are practical. It is not likely that we can accurately estimate such fine lines early on, but the idea of practical limits can be kept in mind so that we seek the needed material but do not press ahead needlessly. If a machine part fails and someone is injured, it is not necessary to investigate the failure beyond the type and nature of the failure and how it contributed to the mishap. It is enough to let metallurgical examination be the province of the manufacturer of the machine or part. The investigative task will be to find how company operation allowed the machine to get into the system, what operational abuses have been involved, what purchasing department deficiencies are indicated, and so on. We will not investigate beyond the things we can control or can hope to correct.

A valued investigative tool is the experience and information gained from past investigations. Through desktop recall of similar events in our own or other systems, we can be directed in our efforts and can correlate the importance of seemingly minor or insignificant findings, which together reveal a major problem area.

Investigative efficiency will be increased by using laptop computers to diagnose electromechanical and electrical equipment failures on the spot, even in remote locations. This will be a great improvement over having to retrieve a system part and send it to a laboratory for the same diagnosis.

Investigations will be speeded through electronic visual media. Travel and per diem costs can be saved by showing a part on a screen to a party thousands of miles away and getting an opinion. Perhaps it will involve playing a tape that needs interpretation or asking face-to-face questions for immediate responses. Much of this will be done without leaving the relative comfort of the computer location. Meetings or trips that take forever to coordinate can be eliminated through electronic viewing and exchange of information.

We do not want to lose track of how an investigation is going. By entering data into our processors as it is gathered, the progress of the investigation can be seen and used to direct us to areas of incomplete information. It is easy to see ourselves entering key elements from witness statements into a processing unit and then inquiring from the computer what the preponderance of information tells us. We can ask the computer to keep rolling costs on the investigation, thus keeping us within financial constraints, even though it may not be the least bit palatable. A related application is information tracking, which enables an investigator to quickly find the status of pending replies to memos and requests for information as well as automatically generating reminders to respond.

Selective Investigation

Now, considering that we do not want to investigate any more than we need to, what about the thought that minor mishaps and incidents that are not really reportable should also be investigated? What may seem like an extravagant use of resources could in reality be a wise investment. If we investigate the near misses, close shaves, and operating inefficiencies in the same spirit of inquiry and concern as we show for a major mishap, the resource investment may be extremely cost-effective. It is far less costly to investigate an event that could have been a serious mishap but by slight chance is not. A day of investigation may replace weeks of work had the event resulted in a mishap.

How do we efficiently investigate, with a minimum of resources, those things that are recordable mishaps in all ways except that they did not result in injury or property damage? We do it selectively and carefully. For example, suppose you are aware that mechanical equipment failures are a problem areas that could cause serious mishaps. Try to find the problem before a mishap results, and do it with the same concern and dedication as if there had been a mishap. Suppose there are too many mechanical equipment failures. Select those to be investigated by careful preliminary analysis or on a random basis, but in proportion to the possible results. This does not mean that you should select only possibly catastrophic events. Certainly they should be considered. But if 80 percent of the problems are with lifting and 1 percent are with fires, you should first selectively look in detail at the mishaps involving lifting. The idea is to look at those things, however mundane, that are costing the most in resources. Lifting would receive special attention until the problem is solved or under control.

When investigating nonreportable mishaps it is easy to get into the "second guessing" business, that is, thinking we know what is most important. We can justify random sampling of problems to be thoroughly investigated, since many unsuspected areas of concern will be found that have not even been considered as a problem.

The Coming Computerization

You may be concerned that a "Big Brother" system will evolve that will take away some of our personal initiative and the personal touch based on our own experiences that make for good investigation. But computers cannot replace investigators. They will lend a hand and enable us to do a more efficient job with better results. The near future does not hold robots going to the field to investigate a mishap. However, the use of robots and computers for field assistance is far from unique. Our end goal is more efficient job operations and assured mission accomplishment. For the private sector, "efficient operations" conserve costly resources, improve operations, and the like. While public agencies, including the military, have budget constraints, they are more mission-oriented, whether it be defense of the country, public service, education, protection, or whatever. Solid, efficient investigation followed by the right corrective action works for either the public or private sector, and high tech helps us do it better.

While face-to-face interaction cannot be replaced in investigation, visual telecommunications devices have untold possibilities. Imagine being in daily face-to-face contact with experts thousands of miles away and using their expertise as if they were with you at a mishap site. Recordings of conversations, graphs, diagrams, and image flows would be available for instant recall. Think of the mishap where a large civil transport aircraft crashed in the Antarctic without survivors. The investigation conditions could not be more severe. Yet we saw the manufacturer's representative in his snow-covered mountainside tent with a light microprocessing unit. He was bouncing signals off a satellite to his home office in Long Beach seeking diagrams of this or that system to aid the field investigation. Perhaps someone comes into the tent from the human factors group and asks for a line to the airline headquarters for data from the pilot's medical records. Some of this did happen at the time, and more was possible. From conservations with the investigators we know how it aided the investigation.

Or consider recent cases where space vehicles have ceased to function (surely considered a mishap). Through careful analysis the causes of malfunction were diagnosed and the vehicles worked back into operation. Still other cases have involved space vehicles that were destroyed and nearly all evidence was thought to be lost. Yet new techniques have made even these difficult and remote mishaps candidates for causal determination.

Analysis

This book is not only a "how to investigate" guide, but also an examination of the analytical process. The distinction between the two is in the eye of the viewer, but it does call for more examination. An excellent investigator may be limited to finding the facts by direction or by ability. Expert analysis of the facts is the next step. The two steps, well done, effectively multiply the overall benefits of the investigation. Finding the facts is only a start. An efficient analysis of the facts leads to firmer and more conclusions. In turn, a good analysis of the conclusions provides more and better recommendations for specific corrective action—far more of them. This can be enhanced by the new investigative aids discussed in this chapter. The result is more than a doubling of efficiency; it increases in at least an algebraic progression. That is, a good investigation more than doubles the number of valid conclusions (we are up to eight times as effective). What leverage! Unfortunately, the concept is based on effective corrective action after the recommendations have been made, but the investigator/analyst has done his share.

Aside from the businesslike processing of facts, recall that there is also room to consider nonfactual information. Facts are essential, but they are not always found. Without tangible information, we must make educated guesses (speculation), to form opinions and judgments on what we know is insufficient evidence (conjecture). This permits us to advance unproven theories (hypothesize) as a reasonable explanation of events that cannot be verified by tests and examination. Disproving a theory is as important as proving one, and so it goes in investigation. Do not, however, spend time chasing theories that stand little chance of being proved.

Retrieving pertinent evidence from a mass of facts and near facts is a selective, analytical process governed by our ability and resources. By what mysterious process do we determine what evidence is pertinent and relevant? "Mysterious" is an adjective often used for things we do not understand, but now that we have the analytical tools, the mystery should be gone. The mystery lies only in which process or combination of processes we will seek our information. This points out that there is no universally accepted method for identifying essential information. You must use the one best suited to the purpose at hand or the guidance given by your company.

With the information gathered, priorities are assigned for the importance of the recommendations and corrective actions. This is done intuitively and through our processing equipment. Artificial intelligence capabilities can allow us to decide which actions give the best cost benefits and most immediate results, considering political environment, operating efficiency, and a score of other items. The tools are there for quickly doing the task. A larger question is whether we can and will take advantage of them.

The need for thorough analysis as a defense against legal action cannot be overlooked. Whether the action is by the courts in products liability or actions of regulatory agencies such as OSHA or the EPA, electronic processing can be invaluable. This protection is not only for the private sector, since government agencies often find themselves in the spotlight for neglect in these areas. In addition, some agencies are legally responsible for their action or lack of action in the mishap arena. It is not for us to judge whether or not the ready recall of mishap data for lawyers interested in litigation or governmental agencies pursuing a Holy Grail is a benefit. When the need arises, recall can be simple, quick, and easy. We press a few buttons and receive a computerized readout. We can take comfort from the advantages we derive from such a system.

The Cost of Progress

Can we afford to use the equipment coming into use? Yes. The problem is in using it to our benefit. Costs of electronic processing and storage of information have steadily declined. The software to support such systems has been the soft underbelly and usually trails hardware capability by 2 to 3 years. When this book first appeared in 1981, closing this gap was a top priority. Now it has been closed, and programs can be developed without the expensive use of programmer/analyst intermediaries. New programming software has eased this problem.

FINAL QUESTIONS

It is likely that in a few years we will be able to plug all the information obtained in an investigation into a microcomputer and get a readout of the causal factors, recommendations, and specific corrective actions that would prevent this and similar mishaps from recurring. This may be accompanied by a neat breakdown of the complete cost of the investigation and the corrective actions. For the level of investigation we now professionally deal with, that time is now. Unfortunately, we sometimes find that there is no logical relationship between systems and humans, thus once again taking us beyond the limits of our knowledge and ability.

What shall we call this task that is both prevention and investigation? Its functions are so complex that if properly done, it calls for specialists in both areas, and the greater expertise will fall to those who specialize in one area or the other. Most safety persons will continue to work in areas of both investigation and prevention, because there is not room for the two separate specialties in their organizations. C. O. Miller long ago coined, in jest, the term "preventigation." Knowing that the arguments of investigation versus prevention will not be settled for many years, I pass it on in jest, unless you accept the challenge posed by "preventigation."

CLOSING

The many associates who have reviewed this revised work have made one thing very clear. There is not nearly enough information in these pages to make a person an expert investigator. At least another two or three volumes are needed. I know those limitations but I expect that those who have come this far will be considerably more enlightened and more capable investigators. Some examples of the resources available to further enlighten you are given in the list of investigation publications in Appendix H.

FOR FURTHER STUDY

The first edition of this book was often used as a college and university text for mishap investigation. It is expected that this revision will also be used for that purpose. Accordingly, some suggested research/term paper topics are listed below for both graduate and undergraduate study levels.

1. Light-Bulb Analysis in Mishap Investigation.
2. Techniques of Arson Investigation.
3. Techniques for the Investigation of Software Failures in Mishaps.
4. Stress as a Mishap Causal Factor.
5. National Economic Benefits of Mishap Investigation.
6. The Validity of Bureau of Labor Statistics in Pinpointing Mishap Causes.
7. OSHA Efforts Toward an Effective Mishap Reporting System.
8. NIOSH Efforts Toward an Effective Mishap Reporting System.
9. New Metals that Defy Analysis for Mishap Investigation Purposes.
10. Use of Robots in Mishap Investigation.
11. Critical Data Required for Mishap Reports for Management Control.
12. Problems for the Company Investigator in Dealing with Toxic Substances and Hazardous Materials Spills.
13. MORT—An Outmoded System.
14. A Profile of the System Safety Development Center (Idaho)—Personnel and Operation.
15. Beyond Management Oversight and Risk Tree.

APPENDIX A

Hazard Sources

Inherent Hazards

Chemical

Corrosiveness
Toxicity
Flammability
Pyrophoricity
Explosiveness
Oxidizability
Photoreactivity
Hydroreactivity
Carcinogenicity
Shock sensitivity

Electrical

Shock
Short circuit
Sparking
Arcing
Explosion
Radiation
Fire
Insulation failure
Overheating

Radiation

Alpha, gamma, and beta
X ray
Infrared and ultraviolet
Radio and microwave

Mechanical

Weight
Speed or acceleration
Stability
Vibration
Rotation
Translation
Reciprocation
Pinch or nip points
Punching and shearing
Sharp edges
Cam action
Stored energy
Entrapment
Impact
Cutting action

Miscellaneous

Noise
Light intensity
Stroboscopic effect
Temperature effect
Pressure, suction
Emissions
Ventilation
Ignition sources
Decomposition
Slipperiness
Moisture
Aging

Human Hazards

Personal

Ignorance
Boredom and loafing
Negligence
Carelessness
Horseplay
Smoking
Alcohol, drugs
Sickness
Exhaustion
Disorientation
Stress
Physical limitations
Cultural background

Human Error

Failure to perform
Incorrect performance
Incorrect supervision
Incorrect training
Overqualification
Poor judgment

Environmental

Weather
Noise
Temperature
Light
Floor texture
Ventilation
Complexity
Comfort conditions
Warnings
Social factors
Psychological factors

APPENDIX B

A Description of Mood-Altering Drugs

Class	Examples	Major Effects
Hard Narcotics	Opium, its derivatives, and related synthetic compounds; for example, heroin, dilaudid, morphine, codeine, demerol	Euphoria, daydreaming, reduced perception of pain, some sedation Withdrawal symptoms for addicted persons include chills, irritability, vomiting
	Cocaine	Euphoria, stimulation, hyperactivity
Psychedelic Drugs	LSD (alias acid), mescaline, STP, DMT, hashish (hash) and marijuana (alias pot, grass, weed, tea, or Maryjane)	Altered perception of self, environment, color, time, space; some may cause hallucinations, panic, or acute psychosis
Psychotropic Drugs		
Stimulants	Amphetamines (alias speed); for example, Dexedrine, Dexamyl, Ritalin, Preludin Other drugs such as caffeine (alias trade name No-Doz) and Elavil	A sense of well-being, lack of fatigue, reduced appetite; sometimes paranoid psychosis
Sedatives	Barbiturates; for example, Secanol, phenobarbitol, Doriden, Sleep-Eze, alcohol, chloralhydrate, and paraldehyde	Depressed central nervous system functions; reduced consciousness; prolonged reaction time

Class	Examples	Major Effects
Tranquilizers	Miltown, Equanil, Librium, Compoz, Atarax, Vitaril, Serpasil, Thorazine, Sporine, Compazine	Calming effect, relaxation, and sense of well-being without appreciably reduced consciousness. However, especially in combination with alcohol and other sedatives, many do reduce consciousness. Side effects of overdose include muscle incoordination.

APPENDIX C

Methodological Approaches and Analytical Methods to Mishap Investigation

A. Methodological Approaches (Benner)

1. **Epidemiological.** Framework for analysis is the agent/host/environment triad and their interaction.
2. **Clinical.** In medicine and psychology, the direct observation of patients and their behavior.
3. **Trend Forecasting.** Extrapolation of historical data to predict trends.
4. **Statistical Inference.** Analysis of data to find the variable(s) most strongly influencing the probability of future occurrences.
5. **Accident Reconstruction.** The identification of historical events and their reconstruction into an ordered description of events.
6. **Simulation.** The examination of a process by simulating it for study.
7. **Behavioral Modeling.** Descriptive modeling from observed human behavior.
8. **Systems Approach.** Definition of a group of interacting components as a system organized into input, operating input, and feedback elements.
9. **Heuristic.** Learning from history independently by individual investigation.
10. **Adversary.** Pitting advocates of two or more opposing viewpoints against each other to resolve the truth, as in litigation.
11. **Scientific.** Gaining understanding of phenomena by observation, hypothesis, testing, and generalization of natural laws and principles.
12. **Kipling.** The faithful servants "who, what, when, where, why, and how."
13. **Sherlock Holmes.** Events sequencing integrated in the investigator's mind.
14. **Engineering.** Comparing operation or construction of an engineered component or system with its designed operation or construction.

15. **Safety Traditional.** Seeking unsafe acts and/or conditions causing mishaps.

B. Methodological Approaches (Ferry)

1. Events sequencing.
2. Known precedent.
3. All cause/multiple cause.
4. Codes, standards, and regulations (CSR's).
5. The four M's of man, machine, media, management (add mission for a fifth).
6. Reenactment.
7. Reconstruction (of wreckage).
8. Simulation.
9. The Hartford EMP approach.
10. Hazard analysis documentation.
11. Inferential conclusions.
12. Program evaluation review technique (PERT).
13. Critical path method (CPM).
14. Failure mode and effect analysis (FEMA).
15. Technique for human error rate prediction (THERP).
16. Fault tree analysis (FTA).
17. Change analysis.
18. Management oversight and risk tree (MORT).
19. Multilinear events sequencing (MES).
20. Technic of operations review (TOR).

C. Other Analytical Techniques

1. Dimensional analysis.
2. Vector analysis.
3. Psychological profiles.
4. Psychological autopsies.
5. Trend analysis.
6. Failure analysis.
7. Sneak circuit analysis.
8. Decision analysis.
9. Scenario modeling.
10. Preliminary hazard analysis.
11. Job safety analysis.

12. Input/output analysis.

13. Time/loss analysis.

D. Other Methodological Approaches

1. ______________________________

2. ______________________________

3. ______________________________

4. ______________________________

5. ______________________________

6. ______________________________

APPENDIX D

MORT and Fault Tree Symbols

	Mort	Fault Tree
	AND gate	AND gate. All inputs required
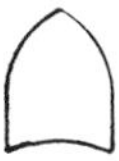	OR gate	OR gate. Any input required
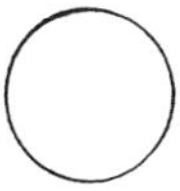	A basic failure event	Primary fault requiring no further development
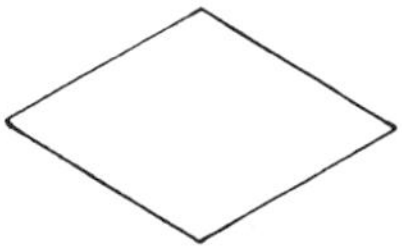	Event where sequence is terminated. Lack of solution or information	Event, usually secondary. Terminated for lack of information
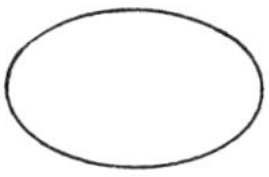	Assumed risk	Restriction in connection with OR, AND gate
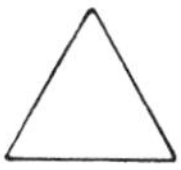	Transfer to another part of tree	Connect or transfer to another part of tree
	Event or fault	Event resulting from combination of events through logic gates
	Event showing satisfactory completion of analysis	None
	Normal event	None

APPENDIX E

Energy Sources

Electrical

Battery banks
Diesel units
High line
Transformers
Wiring
Switchgear
Underground wiring
Cable runs
Service outlets and fittings
Pumps
Motors
Heaters
Power tools
Small equipment

Nuclear (Outside Reactor)

Vaults
Temporary storage areas
Receiving areas
Shipping areas
Casks
Burial grounds
Storage ranks
Canals and basins
Reactor in-tank storage areas
Dollies
Trucks
Hand carry
Cranes
Lifts
Shops
Hot cells
Assembly areas
Inspection areas
Laboratories
Pilot plants

Nuclear (In Reactor)

Reactors
Critical facilities
Subcritical facilities

Kinetic/Linear (In-Plant)

Forklifts
Carts
Dollies
Railroad
Surfaces
Obstructions
Shears
Presses
Crane loads in motion
PV blowdown
Power-assisted driving tools

Kinetic/Linear (Vehicle)

Cars
Trucks
Buses

Kinetic/Rotational

Centrifuges
Motors
Pumps
Cooling tower fans
Cafeteria equipment

Laundry equipment
Gears
Shop equipment (grinders, saws, brushes, etc.)
Floor polishers

PV–KD (Pressure, Tension)

Boilers
Heated surge tanks
Autoclaves
Test loops and facilities
Gas bottles
Pressure vessels
Coiled springs
Stressed members
Gas receivers

MGH (Falls and Drops)

Human effort
Stairs
Bucket and ladder
Trucks
Elevators
Jacks
Scaffolds and ladders
Crane cabs
Pits
Excavations
Elevated doors
Canals
Vessels

MGH (Cranes and Lifts)

Lifts
Cranes
Slings
Hoists

Flammable Materials

Packing materials
Rags
Gasoline (stored and in vehicles)
Oil
Coolant oil
Paint solvent
Diesel fuel
Buildings and contents
Trailers and contents
Grease
Hydrogen (including battery banks)
Gases, other
Spray paint
Solvent vats

Radiation

Canals
Plug storage
Storage areas
Storage buildings
Radioactive sources
Waste and scrap
Contamination
Irradiated experimental and reactor equipment
Electric furnace
Blacklight (e.g., Magnaflux)
Lasers
Medical x rays
Radiography equipment and sources
Welding
Electric arc, other (high-current circuits)
Electron beam
Equipment noise
Ultrasonic cleaners

Thermal Radiation

Furnaces
Boilers
Steam lines
Lab and pilot plant equipment
Sun

Thermal (Except Radiant)

Convection
Heavy metal weld preheat
Exposed steam pipes
Electric heaters
Lead melting pot
Electrical wiring and equipment
Furnaces

Corrosive

Acids
Caustics
"Natural" chemicals (soil, air, water)
Decontaminant solutions

Explosive Pyrophoric

Caps
Primer cord
Dynamite
Power metallurgy
Dusts
Hydrogen (including battery banks/water decomposition)
Gases, other
Nitrates
Peroxides, superoxides

Toxic/Pathogenic

Fluorides
Carbon monoxide
Lead
Ammonia and compounds
Asbestos
Trichloroethylene
Dusts and particulates
Pesticides, herbicides, and insecticides
Bacteria
Beryllium and compounds
Chlorine and compounds
Sandblast, metal plating
Asphyxiation, drowning

APPENDIX F

Questions for MORT Analysis*

The following questions may be used in conjunction with the MORT chart.

Primary Questions

What happened?
Why?
What were the losses (specify the number and type of injuries, property damage, production downtime, product degradation, reduction in employee morale, negative publicity, or any other type of loss involved).
Will there be future undesired events?

S Questions

What were the specific (S) management oversights and omissions that led to the accident?

SA1 Accident (Describe what happened.)
SA2 Was the amelioration LTA (less than adequate)?
*a*1 Was a second accident prevented?
*b*1 Was the plan to prevent a second accident LTA?
*b*2 Was the execution of such a plan LTA?
*c*1 Was practice for execution LTA?
*c*2 Were there personnel or equipment changes that led to poor execution?
*d*1 Was there failure to remedy a personnel or equipment change? Were there task errors that allowed a second accident to occur?
*a*2 Was fire fighting LTA? (Ask questions *b*1–*d*1 above as they apply to fire fighting, that is, was there a plan for fire fighting, was it executed, etc.?)
*a*3 Was there a rescue attempt that was LTA?
*a*4 Was the emergency medical service LTA?
*b*3 Was the first aid LTA?
*b*4 Was transportation LTA?

* Source: G. Driessen.

*c*3 Was there a plan for transportation?
*c*4 Was there notice of the need for transportation?
*c*5 Were personnel and equipment available for transportation?
*c*6 Is the distance a factor? (Is the distance an assumed risk?)
*b*5 Was the treatment LTA?
*a*5 Was the rehabilitation LTA?
*b*6 For persons?
*b*7 For objects (i.e., proper repair after being damaged)?
*a*6 Were relations LTA?
*b*8 Employee relations LTA?
*b*9 Relations with officials LTA?
*b*10 Public relations LTA?
SB1 Did an incident occur? (Describe it.)
Was there a failure to monitor and review the event?
SB2 Were the barriers LTA? (Were there attempts to limit the energy buildup, substitute a safer form of energy, prevent the release, or provide for slow release?)
*a*1 On the energy source?
*a*2 Between the energy source and the persons or objects damaged?
*a*3 On the persons or objects?
*b*1 Were none possible? (Is this an assumed risk?)
*b*2 Was there a failure to provide them where barriers were possible? (Is this an assumed risk?)
*b*3 Did the barrier fail?
*b*4 Was there a barrier available but not used? (Was there a task error?)
*a*4 Was there a failure to separate in time or space the energy and persons, objects?
SB3 Were there persons, objects in the energy channel?
*a*1 Was no evasive action taken?
*b*1 Was none possible? (R5–Is this an assumed risk?)
*a*2 Were the persons, objects functional (i.e., supposed to be there)?
*a*3 Were they nonfunctional?
*b*2 Was control LTA?
*b*3 Was control impractical? (Is this an assumed risk?)
SCI Was there a specific type of unwanted energy flow? (Describe it. Label it unwanted energy flow.) (Was there a failure to monitor this energy flow?)
SDI Is the technical information system less than adequate?
*a*1 Was the technical information system LTA (in regard to unwanted energy flow)?
*b*1 In the area of knowledge:
*c*1 Is there known precedent (i.e., for the prevention of unwanted energy flow)?
*d*1 Is the known precedent available in any written or printed codes, manuals, or recommendations?

*d*2 Is there unwritten precedent (i.e., in the supervisor's regular practice?)
*d*3 Are there any lists of experts to contact in case need should arise?
*d*4 Has any research been directed toward solution of the problem?
*c*2 Is there a known precedent?
*d*5 If not, will there be an accident investigation and analysis of this event?
*d*6 Will there be research?
*b*2 In regard to communications:
*c*3 As applied to internal communications:
*d*7 Is the communication network defined?
*d*8 Does it operate adequately?
*a*2 Are the monitoring systems LTA?
*b*3 Is there routine supervision by management?
*b*4 Is there a search out?
*b*5 Are the accident/incident systems?
*b*6 Are there recorded safety observation (RSO) studies?
*b*7 Are there error-sampling systems?
*b*8 Is there routine hazard potential analysis, hazard inspection, and so on?
*b*9 Is there upstream process auditing?
*b*10 Is there general health monitoring?
*a*3 Is the data reduction system LTA?
*b*11 Is there a priority problem list?
*b*12 Are there summaries, rates, projections, and trend analyses available?
*b*13 Is there any diagnostic statistical analysis performed?
*b*14 Is there depth analysis of special problems?
*a*4 Are there established procedures to control fixes?
*b*15 Are they one-on-one fixes?
*b*16 Is there a system for priority problem fixes?
*b*17 Is there a system for planned change controls?
*b*18 Is there a system for unplanned change fixes?
*b*19 Is there a system for new information use?
*a*5 Is there any independent audit and appraisal?
*a*6 Is there a war room, that is, a centralized location for the display and analysis of hazard and accident data?
SD2 Was the design and plan LTA (that is, in regard to unwanted energy flow **1**)?
*a*1 Were the design planning techniques LTA?
*a*2 Were the organizational and functional responsibilities LTA?
*a*3 Were the interfaces with operations, maintenance, and test organizations LTA?
*a*4 Was the definition of safety criteria LTA? (This includes operations, considerations and availability, materials, fabrication, con-

struction, test, operation, maintenance, and quality assurance requirements.)

*a*5 Was there internal review?
SD3 Was maintenance LTA?
*a*1 Was the maintenance plan LTA?
*b*1 Was there failure to specify a plan?
*c*1 Was maintainability LTA?
*c*2 Was the schedule LTA?
*c*3 Was competence LTA?
*b*2 Was there a failure to analyze failures for causes?
*a*2 Was execution LTA? (Was there task error?)
*b*3 Was there a failure to maintain a P.O. log?
*b*4 Did maintenance cause the failure?
*b*5 Was time LTA?
SD4 Was inspection LTA?
SD5 Was supervision LTA?
*a*1 Was help or training LTA?
*a*2 Was time LTA?
*a*3 Was the job transfer plan LTA?
*a*4 Was there a failure to detect and correct hazards?
*b*1 If there was a failure to detect hazards:
*c*1 Was knowledge LTA (checklists?)?
*c*2 Was the detection plan LTA?
*d*1 Were the logs or schematics LTA?
*d*2 Was the monitor plan LTA?
*d*3 Was there a failure to review changes?
*d*4 Was there a failure to correlate errors?
*c*3 Was time LTA?
*b*2 If there was a failure to correct a hazard:
*c*4 Was interdepartmental coordination LTA?
*c*5 Was there a delay in correction?
*d*5 Was authority LTA?
*d*6 Was budget LTA?
*d*7 Was time LTA?
*c*6 Was the ongoing program (e.g., housekeeping) LTA?
*a*5 Were there performance errors? (Describe unsafe acts.)
*b*3 Were there task errors? (Was there a failure to monitor work errors?)
*c*7 Was there a failure to obtain a work permit?
*c*8 Was job safety analysis (JSA) performed?
*d*8 For repetitive jobs:
*e*1 Was JSA not required?
*e*2 Was JSA required but not done? (Was there a failure to monitor here?)
*d*9 For nonrepetiive jobs:
*e*3 Was the job one with low hazard potential? (Is this an assumed risk?)

*e*4 Was the job one with high hazard potential?
*f*1 Was the prejob analysis LTA?
*f*2 Was there a failure of supervision in this regard?
*g*1 Directly?
*g*2 Correctly?
*c*9 Was JSA LTA?
*d*10 Was knowledge LTA?
*e*5 Were employee suggestions LTA?
*f*3 Is the general employee suggestion system LTA?
*f*4 Are critical incident studies (RSO's) LTA?
*f*5 Was there a failure to use miscellaneous suggestions?
*d*11 Did procedures fail to meet criteria?
*e*6 Was there a failure to correlate with hardware change? (Human factors review.)
*e*7 Was clarity and adequacy of procedures LTA?
*e*8 Was there a failure to verify accuracy?
*e*9 Were the cautions and warnings LTA?
*e*10 Was the specified sequence of events LTA?
*e*11 Were lockouts LTA?
*e*12 Were communications interfaces LTA?
*e*13 Was there a failure to specify personnel environment?
*e*14 Were emergency provisions LTA?
*c*10 Was there a failure to use JSA? (Was there a failure to monitor?)
*d*12 In regard to preparation:
*e*15 As related to selection:
*f*6 Were selection criteria LTA?
*f*7 Were methods of selection LTA?
*e*16 As related to training:
*f*8 Was training totally absent?
*f*9 If training was outside department:
*g*3 Was safety training LTA?
*g*4 Was job content training LTA?
*f*10 If training was inside department, was it LTA?
*f*11 Was there occasion during training for the employee to see correct job performance?
*f*12 Was there a task error even when the employee had seen correct job performance?
*d*13 Were there deviations?
*e*17 Were these normal variability?
*e*18 Were these a result of changes?
*e*19 Was there a failure to observe deviations?
*e*20 Was there a failure to correct deviations?
*f*13 Was there a failure to reinstruct?
*f*14 Was there a failure to enforce regulations?
*d*14 Was employee motivation LTA?

*e***21** Was there schedule pressure?
*e***22** Was there an effort to avoid discomfort?
*e***23** Was there a lack of management concern, vigor, and example?
*e***24** Was job interest building LTA?
*e***25** Was there a conflict in group norms?
*f***15** Was JSA participation LTA?
*f***16** Was innovation diffusion LTA?
*e***26** Was there personal conflict?
*f***17** With the supervisor?
*f***18** With others?
*f***19** Deviant? (**R9**–Is this an assumed risk?)
*e***27** Is the general motivation program LTA?
*c***11** Was the task assignment LTA?
*b***4** Was emergency shutoff LTA? Was there a task error? Was there an effort to prevent a second accident?
*b***5** Were performance errors related to a nontask?
*c***12** Was the nontask peripheral?
*c***13** Was the nontask unrelated?
*c***14** Was the nontask prohibited?
SD**6** Was higher supervision LTA?
SC**2** Were the barriers LTA?
SC**3** Was there an unwanted energy flow?

R Questions

What are the assumed risks? These refer to risks that are known and accepted. They may be risks that are tolerably low (minor), high but impossible to remedy (earthquakes), or simply too expensive to remove. In some cases, risks may be accepted because of the overpowering consideration of other values, for example, the aesthetic value of a waterfall in a national park versus the rare risk of a person's being killed by being drawn over it. The assumed risk questions are specified in other parts of this listing.

G Questions

What are the general (G) management system inadequacies or weaknesses that allowed the accident or incident to occur in the past, and what future undesired events may occur because of them?

GA**1** Is policy LTA?
GA**2** Is the management system implementation LTA?
*a***1** Are the methods, criteria, and analyses LTA?
*a***2** Is line responsibility LTA?
*a***3** Is staff responsibility LTA?

*a***4** Is organization LTA?
*a***5** Are the directives LTA?
*a***6** Is management training and assistance LTA?
*a***7** Are budgets LTA?
*a***8** Are there too many delays?
*a***9** Is accountability LTA?
*a***10** Is vigor and example LTA?
GA**3** Is the risk assessment system LTA?
GB**1** Are the risk assessment goals LTA?
GB**2** Are the risk assessment data analyses LTA?
GB**3** Is the hazard review process LTA?
GC**1** Are the triggers LTA?
GC**2** Is the technical information system LTA? (See SD**1**. These questions are essentially similar here to those on the S branch of the tree and are thus not repeated.)
GC**3** Is the hazard review process undefined?
*a***1** Concepts and requirements.
*b***1** What are the goals and tolerable risks:
*c***1** In relation to safety?
*c***2** In relation to performance?
*b***2** Are the following safety analysis criteria used?
*c***3** A plan?
*c***4** Scaling mechanisms (i.e., a way of expending major effort on major problems and minor effort on minor problems)?
*c***5** Analysis methods? (What are they?)
*c***6** Required alternatives?
*c***7** Specified safety precedence sequences (e.g., priorities for design)?
*c***8** Analyses of environmental impact?
*b***3** Are the following safety requirement criteria used?
*c***9** DOE?
*c***10** OSHA?
*c***11** Other federal codes? (Specify.)
*c***12** State and local codes? (Specify.)
*c***13** Other national codes, standards, and recommendations (Specify.)
*c***14** Internal standards? (Specify.)
*b***4** Was an information search performed?
*c***15** What was the nature of the search?
*c***16** What was the scope of the search?
*b***5** Was a life cycle analysis performed?
*c***17** What was its scope?
*c***18** Is there a requirement for life cycle and failure estimates?
*c***19** Are there required safety factors for extended use?
*b***6** Was there an absence of a hazard review? (Was there a failure to require it or a failure to monitor?)
*A***2** Design and development procedures.

*b*7 In relation to energy control procedures:
*c*20 Are there any unnecessarily exposed hazards?
*c*21 Is there a problem of underdesign?
*c*22 Are the automatic controls LTA?
*c*23 Are the warnings LTA?
*c*24 Are the manual controls LTA?
*c*25 Are there mechanisms for safe energy release?
*c*26 Are the barriers LTA?
*b*8 Has there been an adequate human factors review?
*b*9 Is there an adequate maintenance plan?
*b*10 Is there an adequate inspection plan?
*b*11 What is the arrangement?
*b*12 What is the environment?
*b*13 Are the operability specifications adequate in the areas of:
*c*27 Test and qualification?
*c*28 Supervision?
*c*29 Procedures criteria?
*c*30 Personnel selection?
*c*31 Personnel training and qualification?
*c*32 Personnel motivation?
*c*33 Monitor points?
*c*34 Emergency plans (including amelioration)?
*b*14 Are the change review procedures adequate?
*b*15 Is the disposal plan adequate?
*b*16 Is the independent review technically competent in terms of method and content?
*b*17 Is configurational control adequate?
*b*18 Is documentation adequate?
*b*19 Is there a plan for fast action or expedient cycle?
*b*20 Is the general design process LTA?
*c*35 Are there procedures for code compliance?
*c*36 Are there procedures for use of new codes?
*c*37 Are there procedures for information use?
*c*38 Are there engineering studies to assure compliance with criteria?
*c*39 Has there been any identification of weaknesses and analysis of tradeoffs?
*c*40 Has provision been made for preventive design features?
*c*41 Is there standardization of parts?
*c*42 Is there qualification on nonstandard parts?
*c*43 Are there design descriptions?
*c*44 Is there classification of items according to essentially on safety?
*c*45 Is there identification of items?
*c*46 Are there acceptance criteria?
*c*47 Is there interface control within design processes?
*c*48 Is there development planning?

***c*49** Is there development and qualification testing?
***c*50** Is there test control?
***c*51** Is there development review?
***c*52** Is there failure reporting?
***GB*4** Is the safety program review LTA?
***a*1** Are the ideals defined?
***a*2** Are there satisfactory descriptions and schematics (especially the schematics)?
***a*3** Is there monitoring of the safety program? Audit? Comparison with others?
***a*4** Is the safety program organization adequate in terms of:
***b*1** Scope?
***b*2** Integration (coordination)?
***b*3** Management peer committees?
***a*5** Is the safety program staff adequate?
***a*7** For various operations, are there block-function work schematics available?
***b*4** Are they up-to-date?
***b*5** Do they specify the steps necessary to fulfill each function?
***b*6** Do they contain criteria to know when a job is well done?
***GB*5** Final question. What else?

APPENDIX G

Fire-Associated Temperatures

Material	Melting Point	Remarks
Acetate	887°F	Ignites at 887°F
Aluminum	1220°F	Ignites 1832°F
Benzene		Flash point −4°F
Benzine		Flash point 0°F
Brass	1650°F	
Chromium	3430°F	
Copper	2000°F	
Engine oil		Flashes at 400°F, ignites at 790°F
Ethyl alcohol		Ignites at 680°F
Fuel oil		Flash point 100°F, ignites at 400°F
Gasoline (auto)		Flashes at −50°F, ignites at 750°F
Glass	2600°F	
Hydraulic fluid		Flashes at 200°F, ignites at 640°F
Iron	2795°F	
JP-4		Flashes at −50°F, ignites at 475°F
Magnesium alloy	1250°F	
Neoprene		Blisters at 500°F
Nomex (fire retardant)		Ignites at 2372°F
Nylon fiber	315–420°F	
Nylon tubing	250–300°F	
Paint		Softens at 400°F, blisters at 800°F
Silicone		Blisters at 700°F
Silver	1760°F	
Soot		Will not adhere to surface over 700°F
Stainless steel	2700°F	Starts discoloring at 850°F
Teflon	630°F	Ignites at 1050°F
Tin	449°F	
Titanium	3140°F	Scales form at 110°F
Tungsten	6170°F	
Wire bundles		Outer braid brittle at 500°F
Wood		Ignites at 400–500°F
Zinc chromate		Tans at 450°F, browns at 600°F

APPENDIX H

Investigation Publications

1 *Accident Investigation: A New Approach,* National Safety Council, Chicago, 1983, 44 pages. 0-87912-036-3.

2 *Automotive Engineering and Litigation,* Vol. 1, George A. Peters and Barbara J. Peters, Eds., Garland Law Publishing, New York, 1984. KF1207A8A97.

3 Bodurtha, Frank T., *Industrial Explosion, Prevention and Protection,* McGraw-Hill, New York, 1979. T55.3 E96 B6.

4 Brannigan, F. L., et al., *Fire Investigation Handbook* (PB81-113482), NTIS, Washington, DC, 1980. TH9180 B7.

5 Collins, James C., *Accident Reconstruction,* Charles C. Thomas, Springfield, IL, 1979. HV8079.55 C94.

6 Ellis, Glenn, *Air Crash Investigation of General Aviation Aircraft,* Capstain Publications, Greybill, WY, 1984. TL553.5 E45.

7 Epstein, Richard A., *Modern Products Liability Law,* Quorum Books, Westport, CT, 1980. KF1296 E67.

8 Ferry, Ted S., *Readings in Accident Investigations,* Charles C. Thomas, Springfield, IL, 1984.

9 Paul L. Tung, Ed., *Fracture and Failure: Analysis, Mechanisms and Application,* American Society for Metals, Metals Park, OH, 1980. TS227.8 F73.

10 Human Factors Group, Eastman Kodak Company, *Ergonomic Design for People at Work,* Vol. 1, Elizabeth Eggleston, Ed., Lifetime Learning Publications, Belmont, CA, 1983.

11 Hurt, H. H., *Motorcycle Accident Cause Factors and Identification of Counter Measures,* Vol. 1, National Highway Traffic Safety Administration, Washington, DC, 1981. TL440.5 H8.

12 Kirk, Paul L., *Kirk's Fire Investigation,* Wiley, New York, 1983.

13 Kolb, John, and Steven S. Ross, *Product Safety and Liability: A Desk Reference,* McGraw-Hill, New York, 1980. TS175.K64.

14 Limpert, Rudolph, *Vehicle System Components: Design and Safety,* Wiley, New York, 1982.

15 Mazer, William M., *Electrical Accident Investigation Handbook,* Electrodata, Inc., Glen Echo, MD, 1982. 0-943890-03-09. (subscription, two volumes, supplements and revisions).

16 *NATO Symposium on Human Detection and Diagnosis of System Failures,* Plenum, New York, 1980. TA169.5 N37.

17 Perrow, Charles, *Normal Accidents: Living with High-Risk Technologies,* Basic Books, New York, 1984. T54.P47.

18 Petroski, Henry, *To Engineer Is Human,* St. Martins Press, New York, 1985. TA174.P747.

19 *Product Liability Handbook,* Society of the Plastics Industry, New York, 1982. KF1296 P65.

20 *Safety Law: A Legal Reference for the Safety Professional,* George Peters, Ed. American Society of Safety Engineers, Park Ridge, IL, 1983.

21 Smith, Al J., Jr., *Managing Hazardous Substances Accidents,* McGraw-Hill, New York, 1981. T55.3 H3S64.

22 Smith, Charles O., *Product Liability: Are You Vulnerable?*, Prentice-Hall, Englewood Cliffs, NJ, 1981. KF1296 S6.

23 Terrien, Ernest J., *Hazardous Materials and Natural Disaster Emergencies,* Technoic Publishing Co., Lancaster, PA, 1984. 84-51388.

24 Waddell, William C., *Overcoming Murphey's Law,* AMCOM, New York, 1981, HD31 W235.

25 Weinstein, A. S., *Products Liability and the Reasonably Safe Product: A Guide for Management, Design and Marketing,* Wiley, New York, 1978. KF1296 P77.

26 Woodson, Wesley E., *Human Factors Design Handbook: Information and Guidelines for The Design of Systems, Facilities and Equipment,* McGraw-Hill, New York, 1981. TA166 W57.

ADDITIONAL REFERENCES

Index